An Introduction to MATLAB® Programming and Numerical Methods for Engineers

An Introduction
to MATLAB® Programming
and Numerical Methods
for Engineers

Timmy Siauw

Alexandre M. Bayen

AMSTERDAM • BOSTON • HEIDELBERG • LONDON
NEW YORK • OXFORD • PARIS • SAN DIEGO
SAN FRANCISCO • SINGAPORE • SYDNEY • TOKYO
Academic Press is an imprint of Elsevier

Academic Press is an imprint of Elsevier
The Boulevard, Langford Lane, Kidlington, Oxford, OX5 1GB
225 Wyman Street, Waltham, MA 02451, USA
32 Jamestown Road, London NW1 7BY, UK
Radarweg 29, PO Box 211, 1000 AE Amsterdam, The Netherlands
525 B Street, Suite 1900, San Diego, CA 92101-4495, USA

Library of Congress Cataloging-in-Publication Data
Siauw, Timmy.
  An introduction to MATLAB programming and numerical methods for engineers / Timmy Siauw, Alexandre M. Bayen.
    pages cm
  Includes bibliographical references and index.
  ISBN 978-0-12-420228-3 (paperback)
  1. MATLAB. 2. Numerical analysis—Data processing. 3. Engineering—Data processing. 4. Engineering mathematics.
I. Bayen, Alexandre M. II. Title.
  QA297.S4785 2014
  518.0285'53—dc23

                                        2014010755

British Library Cataloguing in Publication Data
A catalogue record for this book is available from the British Library

ISBN: 978-0-12-420228-3

For information on all Academic Press publications
visit our website at **store.elsevier.com**

Printed and bound in the United States

15 16 17 18 19   10 9 8 7 6 5 4 3 2 1

Working together
to grow libraries in
developing countries

www.elsevier.com • www.bookaid.org

*To the students of UC Berkeley's E7 class: past,
present, and future.*

*A ma mère Catherine Bayen, avec reconnaissance
pour son entier dévouement á mon éducation.*

Contents

PART 1   INTRODUCTION TO PROGRAMMING FOR ENGINEERS

PART 2   INTRODUCTION TO NUMERICAL METHODS

Preface

Purpose

Because programming has become an essential component of engineering, medicine, media, business, finance, and many other fields, it is important for engineers and practitioners to have a basic foundation in programming to be competitive. This book introduces programming to students from a wide range of backgrounds and gives them programming and mathematical tools that will be useful to them throughout their careers.

For the most part, this book follows the standard material taught at the University of California, Berkeley, in the class *E7: Introduction to computer programming for scientists and engineers*. This class is taken by most engineering freshmen in the College of Engineering and by undergraduate students from other disciplines, including physics, biology, and cognitive science. The course has two fundamental goals:

- Teach MATLAB programming to engineering students who do not have prior exposure to programming.
- Introduce a variety of numerical analysis tools that are useful for solving engineering problems.

These two goals are reflected in the two parts of this book:

- Introduction to Programming for Engineers
- Introduction to Numerical Methods

Because this book covers such a wide range of topics, no topic is covered in great depth. In fact, each chapter is designed to be covered in at most two lecture hours, even though there are entire semester courses dedicated to these same chapters. Rather than depth, this book is intended to give students a wide breadth of programming knowledge and mathematical vocabulary on which they can expand.

We believe that just like learning a new foreign language, learning to program can be fun and illuminating. We hope that as you journey through this book, you will agree.

Prerequisites

This book is designed to introduce programming and numerical methods to students who have *absolutely no* prior experience with programming, which we hope is reflected in the pace, tone, and content of the text. For the purpose of programming, we assume the reader has the following prerequisite knowledge:

- Understanding of the computer monitor and keyboard/mouse input devices
- Understanding of the folder structure used to store files in most operating systems

For the mathematical portions of the text, we assume the reader has the following prerequisite knowledge:

- High school level algebra and trigonometry
- Introductory, college-level calculus

That's it! Anything in the text that assumes more than this is our mistake, and we apologize in advance for any instances that might pop up.

Organization

Part I teaches the fundamental concepts of programming. Chapter 1 introduces the reader to MATLAB and MATLAB's basic interface. Chapters 2 through 6 teach the fundamentals of programming. Proficiency in the material from these chapters will allow you to program almost anything you imagine. Chapter 7 provides theory that characterizes computer programs based on how fast they run, and Chapter 8 gives insights into how computers represent numbers and their effect on arithmetic. Chapter 9 explains how to store data in the long term and how to make results from MATLAB useful outside of MATLAB (i.e., for other programs). Chapter 10 provides useful tips on good programming practices to limit mistakes from popping up in your code, and tells you how to find them when they do. Finally, Chapter 11 introduces MATLAB's graphical features that allow you to produce plots and charts, which is arguably one of MATLAB's most useful features for engineers.

Part 2 gives an overview of a variety of numerical methods that are useful for engineers. Chapter 12 gives a crash course in linear algebra. Although theoretical in nature, linear algebra is the single most critical concept for understanding many advanced engineering topics. Chapter 13 is about regression, a method of fitting theoretical models to observed data. Chapter 14 is about inferring the value of a function between observed data points, a framework known as interpolation. Chapter 15 introduces the idea of approximating functions with sums of polynomials, which can be useful for simplifying complicated functions. Chapter 16 teaches two algorithms for finding roots of functions. That is, find an x such that $f(x) = 0$, where f is a function. Chapters 17 and 18 cover methods of approximating the derivative and integral of a function, respectively. Finally, Chapter 19 introduces a mathematical model type called ordinary differential equations, and presents several methods of finding solutions to them.

What's Missing?

Since no prior programming knowledge is assumed for this book, it is important to state clearly what is *not* taught in this text. All the programming concepts in this text fall under a style of programming called Procedural Programming, which basically means building computer programs that work step by step to complete a task. This is a fundamentally different approach than Object-Oriented Programming, which emphasizes building concepts as computational objects that help programmers keep track of large projects.

Object-Oriented Programming is most effective when used for very large programming projects, usually projects that involve multiple programmers working together. We have omitted it from this text primarily because there is insufficient time in a single semester to teach Procedural Program well *and* give exercise problems of a size and scope that would demonstrate the effectiveness of Object-Oriented Programming. That being said, Object-Oriented Programming is a very powerful programming paradigm, and we hope that you will explore it in the future.

As suggested earlier, this text does not provide mathematically rigorous definitions of what the methods presented are or why they are effective. There are some mathematical derivations but no mathematical proofs. Our primary motivation for this text is to give you a foundation of programming and mathematical tools that you can use to solve problems. We leave rigor and depth to your future courses and more advanced textbooks.

How to Read this Book

Learning to program is all about practice, practice, practice. Just like learning a new language, there is no way you will learn to program well without engaging with the material, internalizing it, and putting it into constant use.

As you go through the text, you should ideally have MATLAB open in front of you, ready to try out any and all of the numerous examples that are provided. Go *slowly*. Taking the time to really understand what MATLAB is doing in every example will pay large dividends compared to "powering through" the text like a novel.

In terms of the text itself, Chapters 1 through 5 should be read and understood first since they cover the fundamentals of programming. Chapters 6 through 10 can be covered in any order. Chapter 11 on plotting is a must-read but can be covered any time. In Part II, Chapter 12 should be read first since subsequent chapters rely on linear algebraic concepts. The remaining chapters can be read in any order. However, it will be helpful to read Chapters 17 and 18 before Chapter 19.

Throughout the text, there will be words written in **boldface**. When you encounter one of these words, you should take the time to commit the word to memory and understand its meaning in the context of the material being presented.

To keep the text from running on, we punctuate the material with smaller blocks. Following is a description of each kind of block.

TRY IT! This is the most common block in the text. It will usually have a short description of a problem and/or an activity. We strongly recommend that you actually try all of these in MATLAB.

TIP! This block gives some advice that we think will make programming easier for you. However, the blocks do not contain any new material that is essential for understanding the key concepts of the text.

EXAMPLE: These sections are concrete examples of new concepts. They are designed to help you think about new concepts. However, they do not necessarily need to be tried.

WARNING! Learning to program can have many pitfalls. These sections contain information that will help you avoid confusion, building bad habits, or misunderstanding key concepts.

WHAT IS HAPPENING? These sections follow MATLAB in scrutinizing detail to help you understanding what goes on when MATLAB executes programs.

CONSTRUCTION: In programming there are standard architectures that are reserved to perform common and important tasks. These sections outline these architectures and how to use them.

There are four sections to end every chapter. The **Summary** section gives a list of the main points of the chapter. These points should be intuitive to you by the end of the chapter. The **Vocabulary** section is a list of new words presented in the chapter. It is important to know how these words are defined in the context of this book since they will be essential for learning concepts later. The **Functions and Operators** section lists new tools and functions introduced during the chapter that will be useful for your programs and for the exercise problems. Be sure to understand what these tools do and how they are used. The **Problems** section gives exercises that will reinforce concepts from the chapter. There are five types of exercise problems and they are each denoted by their own symbol:

The ✎ symbol denotes a problem that you should work out on paper.

The 🧠 symbol denotes a problem that gives you something to think about.

The ≫ symbol denotes a problem that you should try at the command prompt.

The .m symbol denotes a problem for which you need to write a program.

The ● symbol denotes a problem that requires you to intentionally generate an error.

These problems are designed to familiarize you with common mistakes made while programming so that you can readily fix them.

As one final note, one of the main criticisms of MATLAB is that there are too many ways of doing the same thing. Although at first this can seem like a useful feature, it can make learning MATLAB confusing or overload you with possibilities when the task is actually straightforward. This book presents a single way of performing a task to provide structure for your learning experience and to keep you from being inundated by extraneous information. You may discover solutions that differ from the text's solutions but solve the problem just the same or even better! We encourage you to find these alternative methods, and leave it up to experience and your own judgement to decide which way is better.

We hope you enjoy the book!

MATLAB Version

This book was written using MATLAB R2008a. As MATLAB is constantly under development, some features may be added, removed, or changed in the MATLAB version on your computer.

Acknowledgments

This book would never have been written without the help of colleagues, teams of Graduate Student Instructors (GSI), graders, and administrative staff members who helped us through the challenging process of teaching E7 to several hundreds of students each semester at UC Berkeley. Furthermore, this book would never have reached completion without the help of the students who had the patience to read the book and give us their feedback. In the process of teaching E7 numerous times, we have interacted with thousands of students, dozens of GSIs and graders, and a dozen colleagues and administrators, and we apologize to the ones we will inevitably forget given the number of people involved.

We are extremely grateful for guidance from our colleagues Professors Panos Papadopoulos, Roberto Horowitz, Michael Frenklach, Andy Packard, Tad Patzek, Jamie Rector, Raja Sengupta, Mike Cassidy, and Samer Madanat. We owe thanks particularly to Professors Roberto Horowitz, Andy Packard, Sanjay Govindjee, and Tad Patzek for sharing the material they used for the class, which contributed to the material in this book. We also thank Professors Rob Harley and Sanjay Govindjee for using a draft of this book during the semesters they taught E7 and giving us feedback that helped improve the manuscript.

The smooth running of the semester course gave the authors the time and energy to produce this book. Managing the course was greatly facilitated by numerous administrative staff members who bore much of the logistic load. We are particularly grateful to Joan Chamberlain, Shelley Okimoto, Jenna Tower, and Donna Craig. Civil and Environmental Engineering Vice Chair Bill Nazaroff deserves particular recognition for assigning the second author to teach the class in 2011. Without this assignment the two authors of this book would not have had an opportunity to work together and write this book.

E7 is notoriously the hardest class to teach at UC Berkeley in the College of Engineering. However, it continued to run smoothly over the many semesters we learned to teach this class, mainly due to the help of the talented GSIs we had the pleasure of working with. During the years the co-authors taught the class, a series of legendary head GSIs have contributed to shaping the class and making it a meaningful experience for students. In particular, Scott Payne, James Lew, Claire Saint-Pierre, Kristen Parish, Brian McDonald, and Travis Walter have in their respective roles led a team of dedicated GSI to exceed expectations. The GSI and grader team during the Spring of 2011 greatly influenced the material of this book. For their contribution during that critical semester, we thank Jon Beard, Leah Anderson, Marc Lipoff, Sebastien Blandin, Sam Chiu, Rob Hansen, Jiangchuan Huang, Brad Adams, Ryan Swick, Pranthik Samal, Matthieu Lewandowski, and Romain Bourcier.

We are also grateful to Claire Johnson and Katherine Mellis for finding errors in the text and helping us incorporate edits into the manuscript.

Finally, we are indebted to the students for their patience with us and their thorough reading of the material. Having seen thousands of them through the years, we are sorry to only be able to mention a few for their extraordinary feedback and performance: Gurshamnjot Singh, Sabrina Nicolle Atienza, Yi Lu, Nicole Schauser, Harrison Lee, Don Mai, Robin Parrish, and Mara Minner.

List of Figures

Introduction
to Programming
for Engineers

MATLAB® Basics

CHAPTER OUTLINE

Motivation

This chapter gets you started with MATLAB, using it as a calculator. As you will see, MATLAB has a large library of built-in mathematical functions that you can use to perform any operation available on a scientific or graphing calculator. At the end of this chapter, you should be familiar with the MATLAB environment, how to execute commands to MATLAB, and MATLAB's basic mathematical functionalities.

1.1 Getting Started with the MATLAB® Environment

Once MATLAB is installed on your computer, you should see a shortcut on the desktop that looks like

. Double-clicking the shortcut icon will open the MATLAB environment shown in Figure 1.1.

The **MATLAB environment** is a text-based visualization tool that allows you to interact with MATLAB. The MATLAB environment consists of the current directory as well as four windows: the command window, the current directory window, the workspace window, and the command history window. They are shown in Figure 1.1.

The **current directory** is the folder in your computer where files will be saved and where the files you will have direct access to are stored. You can change the current directory by clicking the down arrow or the button with an ellipsis symbol (…). The current directory will be explained in greater detail in Chapter 3 on Functions. The **command window** is the window in the MATLAB environment where commands are executed and MATLAB's responses are displayed. The **command prompt** is where you can type your commands in the command window and is denoted by the symbol ». When you see this symbol in the text or in examples, it means that the action is taking place at the command window. The **current directory window** is the window in the MATLAB environment that lists all the files currently

An Introduction to MATLAB® Programming and Numerical Methods. http://dx.doi.org/10.1016/B978-0-12-420228-3.00001-4

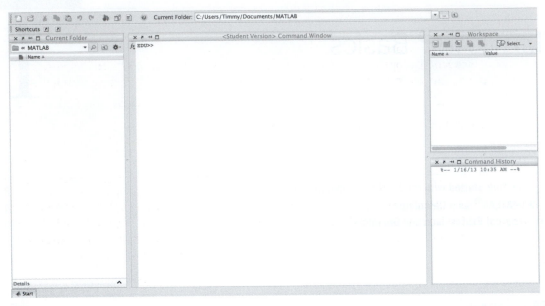

FIGURE 1.1

The MATLAB environment.

stored in the current directory. The **workspace window** is the window in the MATLAB environment that lists all the variables currently being used in the workspace. The details of the workspace will also be explained in Chapter 3 on Functions. The **command history window** is the window in the MATLAB environment that lists all the previous commands entered at the command prompt, which is helpful for recalling work that was done in a previous session.

You can rearrange the size and shape of the windows by clicking and holding the mouse cursor on the borders of the windows, and then dragging them to a location that suits you better. You can also change the location of the windows by clicking and holding the mouse cursor on the window title bar, and dragging it to a different location. If you wish to remove a window, click the X in the upper right corner of the window. To put it back, click Desktop in the menu bar and then click the window you want.

TIP! You can change the background color as well as the font color and style to suit your personal preference. The options can be changed in the File → Preferences menu in the upper left hand corner of the MATLAB environment.

1.2 MATLAB® as a Calculator

We will introduce you to MATLAB by demonstrating features found in any standard graphing calculator. An **arithmetic operation** is either addition, subtraction, multiplication, division, or powers between two numbers. An **arithmetic operator** is a symbol that MATLAB has reserved to mean one of the aforementioned operations. These symbols are + for addition, − for subtraction, ∗ for multiplication, / for division, and ^ for exponentiation.

We say an instruction or operation is **executed** when it is resolved by the computer. An instruction is executed at the command prompt by typing it where you see the » symbol and then pressing Enter.

TRY IT! Compute the sum of 1 and 2.

```
>> 1 + 2
ans = 3
```

An **order of operations** is a standard order of precedence that different operations have in relationship to one another. MATLAB utilizes the same order of operations that you learned in grade school. Powers are executed before multiplication and division, which are executed before addition and subtraction. Parentheses, (), can also be used in MATLAB to supercede the standard order of operations.

TRY IT! Compute $\frac{3*4}{2^2+4/2}$.

```
>> (3*4)/(2^2 + 4/2)
ans = 2
```

TIP! You may have noticed `ans` is the resulting value of the last operation executed. You can use `ans` to break up complicated expressions into simpler commands.

TRY IT! Compute 3 divided by 4, then multiply the result by 2, and then raise the result to the 3rd power.

```
>> 3/4
ans = 0.75
>> ans*2
ans = 1.5
>> ans^3
ans = 3.3750
```

MATLAB has many basic arithmetic functions like `sin`, `cos`, `tan`, `asin`, `acos`, `atan`, `exp`, `log`, `log10`, and `sqrt`. The inputs to these mathematical functions are always placed inside of parentheses that are connected to the function name. For trigonometric functions, it is useful to have the value of π easily available. You can call this value at any time by typing » `pi` in the command prompt. Note that the value of π is stored in MATLAB to 16 digits.

TRY IT! Find the square root of 4.

```
>> sqrt(4)
ans = 2
```

TRY IT! Compute the sin $\left(\frac{\pi}{2}\right)$.

```
>> sin(pi/2)
ans = 1
```

TIP! Sometimes you may wish to view more or less digits than MATLAB's default setting of four decimal places. There are many different number viewing options in MATLAB but for the purposes of this text, we will restrict these options to "short," "long," and "bank" unless you are specifically told otherwise. The short format is MATLAB's default setting. It displays all numbers to four significant figures. The long format displays the maximum number of digits that MATLAB can store, which is 16. The bank format displays exactly two.
You can change the formatting by typing of the following commands:

```
>> format short
>> format long
>> format bank
```
Note that this changes only how the numbers are displayed; it does not alter the actual value being used.

TRY IT! Call MATLAB's stored value for π using format long, format bank, and format short.

```
>> format long
>> pi
ans =
    3.141592653589793
>> format bank

>> pi
ans =
            3.14
>> format short
>> pi
ans =
    3.1416
```

MATLAB will compose functions as you would expect, with the innermost function being executed first. The same holds true for function calls that are composed with arithmetic operations.

TRY IT! Compute $e^{\log 10}$.

```
>> exp(log(10))
ans = 10
```

TRY IT! Compute $e^{\frac{3}{4}}$.

```
>> exp(3/4)
>> ans = 2.1170
```
Note that the `log` function in MATLAB is $\log_e$, or the natural logarithm. It is not $\log_{10}$. If you want to use $\log_{10}$, you need to use `log10`.

TIP! Using the UP ARROW in the command prompt recalls previous commands that were executed. If you accidentally type a command incorrectly, you can use the UP ARROW to recall it, and then edit it instead of retyping the entire line.

The **help function** is a command that can be used to view the description of any function in MATLAB. You can call the help function by typing » `help` at the command prompt and then the name of the function. If you see a function you are unfamiliar with, it is good practice to use the help function before asking your instructors what a specific function does. At the end of every chapter in this book is a section called "Functions and Operators," which lists the new functions and operations presented in the chapter. If you are uncertain what these functions do, use the help function to learn about them.

WARNING! For some functions, the help file can be extremely complicated and wordy, even for simple functions. In these cases, do not be afraid to ask your instructor for help.

TRY IT! Use the `help` function to find the definition of the `factorial` function.

```
>> help factorial
FACTORIAL Factorial function.
FACTORIAL(N) for scalar N, is the product of all the integers from 1 to N,
i.e. prod(1:N). When N is an N—D matrix, FACTORIAL(N) is the factorial for
each element of N.  Since double precision numbers only have about
15 digits, the answer is only accurate for N <= 21. For larger N,
the answer will have the correct order of magnitude, and is accurate for
the first 15 digits.
```

TIP! Use the `format compact` command to reformat text so that you have only a single space between commands instead of the default setting of double space. You can change the spacing format using the command » `format compact`; to change it back, use » `format loose`.

MATLAB can handle the expression 1/0, which is infinity. Note that MATLAB will return 0/0 as "not a number" or NaN. You can type `Inf` at the command prompt to denote infinity or NaN to denote something that is not a number that you wish to be handled as a number. If this is confusing, this distinction can be skipped for now; it will be explained more clearly when it becomes important. Finally, MATLAB can also handle the imaginary number, i, which is $\sqrt{-1}$. You can type » `i` to recall the stored value of i just like π.

TRY IT! Compute $1/0$, $1/\infty$, and $\infty \cdot 2$ to verify that MATLAB handles infinity as you would expect.

```
>> 1/0
ans =
    Inf
>> 1/Inf
ans =
     0
>> Inf * 2
ans =
    Inf
```

TRY IT! Compute ∞/∞.

```
>> Inf/Inf
ans =
    NaN
```

TRY IT! Verify that MATLAB's stored value for `i` squares to -1.

```
>> i^2
ans =
    -1
```

> **TRY IT!** Compute the imaginary sum $2 + 5i$.
>
> ```
> >> 2 + 5*i
> ans =
>      2 + 5i
> ```

MATLAB can also handle scientific notation using the letter e between two numbers. For example, » `1e6` is $1 \times 10^6 = 1000000$ and » `1e-3` is $1 \times 10^{-3} = 0.001$.

> **TRY IT!** Compute the number of seconds in 3 years using scientific notation.
>
> ```
> >> 3e0*3.65e2*2.4e1*3.6e3
> ans =
>      94608000
> ```

1.3 Logical Expressions and Operators

A **logical expression** is a statement that can either be true or false. For example, $a < b$ is a logical expression. It can be true or false depending on what values of a and b are given. Note that this differs from a **mathematical expression** which denotes a truth statement. In the previous example, the mathematical expression $a < b$ means that a is less than b, and values of a and b where $a \geq b$ are not permitted. Logical expressions form the basis of computing, so for the purposes of this book, all statements are assumed to be logical rather than mathematical unless otherwise indicated.

In MATLAB, a logical expression that is true will compute to the value "TRUE." A false expression will compute to the value "FALSE." For the purpose of this book, "TRUE" is equivalent to 1, and "FALSE" is equivalent to 0. Distinguishing between the numbers 1 and 0 and the logical values "TRUE" and "FALSE" is beyond the scope of this book, but it is covered in more advanced books on computing. Logical expressions are used to pose questions to MATLAB. For example, "$3 < 4$" is equivalent to, "Is 3 less than 4?" Since this statement is true, MATLAB will compute it as 1. However, $3 > 4$ is false, therefore MATLAB will compute it as 0.

Comparison operators compare the value of two numbers, and they are used to build logical expressions. MATLAB reserves the symbols $>$, $>=$, $<$, $<=$, $\sim=$, $==$, to denote "greater than," "greater than or equal," "less than," "less than or equal," "not equal," and "equal," respectively.

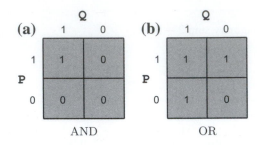

FIGURE 1.2

Truth tables for the logical AND and OR.

> **TRY IT!** Compute the logical expression for "Is 5 equal to 4?" and "Is 2 smaller than 3?"
>
> ```
> >> 5 == 4
> ans =
>      0
> >> 2 < 3
> ans =
>      1
> ```

Logical operators are operations between two logical expressions that, for the sake of discussion, we call P and Q. The fundamental logical operators we will use herein are **AND**, **OR**, and **NOT**, which in MATLAB are denoted by &&, ||, and $\sim$, respectively. There are other logical operators, but they are equivalent to combinations of these three operators. P AND Q is true only if P and Q are *both* true. P OR Q is true if either P or Q is true or if both P and Q are true. It is important to note that OR in MATLAB is "inclusive" OR, meaning it is true if both P and Q are true. In contrast, "exclusive" OR or **XOR** is true if either P or Q is true but false if both P and Q are true. If P is true, then NOT P is false, and if P is false, then NOT P is true.

The **truth table** of a logical operator or expression gives the result of every truth combination of P and Q. The truth tables for AND and OR are given in Figure 1.2.

> **TRY IT!** Assuming P is true, use MATLAB to determine if the expression (P AND NOT(Q)) OR (P AND Q) is always true regardless of whether or not Q is true. Logically, can you see why this is the case?
> First, assume Q is true:
>
> ```
> >> (1&&~1) || (1&&1)
> ans = 1
> ```

Now assume Q is false:

```
>> (1&&~0) || (1&&0)
ans = 1
```

Just as with arithmetic operators, logical operators have an order of operations relative to each other and in relation to arithmetic operators. All arithmetic operations will be executed before comparison operations, which will be executed before logical operations. Parentheses can be used to change the order of operations.

TRY IT! Compute $(1 + 3) > (2 + 5)$.

```
>> 1 + 3 > 2 + 5
ans = 0
```

TIP! Even when the order of operations is known, it is usually helpful for you and those reading your code to use parentheses to make your intentions clearer. In the preceding example `(1 + 3) > (2 + 5)` is clearer than `1 + 3 > 2 + 5`.

WARNING! In MATLAB's implementation of logic, 1 is used to denote true and 0 for false. However, 1 and 0 are still numbers. Therefore, MATLAB will allow abuses such as `>> (3>2) + (5>4)`, which will resolve to 2.

WARNING! Although in formal logic, 1 is used to denote true and 0 to denote false, MATLAB slightly abuses notation and it will take any number not equal to 0 to mean true when used in a logical operation. For example, `3 && 1` will compute to true. Do not utilize this feature of MATLAB. Always use 1 to denote a true statement.

TRY IT! A fortnight is a length of time consisting of 14 days. Use a logical expression to determine if there are more than 100,000 seconds in a fortnight.

```
>> (14*24*60*60) > 100000
ans = 1
```

Summary

1. You can interact with MATLAB through the MATLAB environment.
2. MATLAB can be used as a calculator. It has all the functions and arithmetic operations commonly used with a scientific calculator.
3. You can also use MATLAB to perform logical operations.

Vocabulary

AND	current directory	mathematic expression
arithmetic operation	current directory window	NOT
arithmetic operator	execute	order of operations
command window	help function	OR
command prompt	logical expression	truth table
command history window	logical operator	workspace window
comparison operator	MATLAB environment	

Functions and Operators

+	~=	log
-	==	log10
*	&&	help
/	||	format short
^	~	format long
()	ans	format bank
>	pi	NaN
>=	sin	Inf
<	cos	i
<=	exp	

Problems

⊠ **1.** Remove the command history window from the MATLAB environment and then retrieve it.

⊠ **2.** Resize the command prompt so that it takes up less than half of the total MATLAB environment space.

▷▷ **3.** Change the background of the MATLAB environment to black and the font color to orange.

▷▷ **4.** Change the current directory to any folder other than the current default working directory, and then change it back to the default directory.

▷▷ **5.** Type » `travel` into the command prompt. This program tries to solve the Traveling Salesman Problem.

▷▷ **6.** Type » `filterguitar` into the command prompt. This program simulates the sound of a guitar using mathematical and computational methods.

▷▷ **7.** Type » `lorenz` into the command prompt. The Lorenz Attractor is a mathematical model originally formulated to simulate atmospheric weather patterns. However, it has some surprising results that eventually led to the field of Chaos Theory. This program displays the simulation results for different initial conditions.

▷▷ **8.** Compute the area of a triangle with base 10 and height 12. Recall that the area of a triangle is half the base times the height.

▷▷ **9.** Compute the surface area and volume of a cylinder with radius 5 and height 3.

▷▷ **10.** Compute the slope between the points $(3, 4)$ and $(5, 9)$. Recall that the slope between points (x_1, y_1) and (x_2, y_2) is $\frac{y_2-y_1}{x_2-x_1}$.

▷▷ **11.** Compute the distance between the points $(3, 4)$ and $(5, 9)$. Recall that the distance between points in two dimensions is $\sqrt{(x_2 - x_1)^2 + (y_2 - y_1)^2}$.

▷▷ **12.** Use MATLAB's `factorial` function to compute 6!

▷▷ **13.** A year is considered to be 365 days long. However, a more exact figure is 365.24 days. As a consequence, if we held to the standard 365-day year, we would gradually lose that fraction of the day over time, and seasons and other astronomical events would not occur as expected. A leap year is a year that has an extra day, February 29, to keep the timescale on track. Leap years occur on years that are exactly divisible by 4, unless it is exactly divisible by 100, unless it is divisible by 400. For example, the year 2004 is a leap year, the year 1900 is not a leap year, and the year 2000 is a leap year.

Compute the number of leap years between the years 1500 and 2010.

▷▷ **14.** A very powerful approximation for π was developed by a brilliant mathematician named Srinivasa Ramanujan. The approximation is the following:

$$\frac{1}{\pi} \approx \frac{2\sqrt{2}}{9801} \sum_{k=0}^{N} \frac{(4k)!(1103 + 26390k)}{(k!)^4 396^{4k}}.$$

Use Ramanujan's formula for $N = 0$ and $N = 1$ to approximate π. Be sure to use format long. Compare your approximation with MATLAB's stored value for `pi`. Hint: $0! = 1$ by definition.

▷▷ **15.** The hyperbolic sin or sinh is defined in terms of exponentials as $\sinh(x) = \frac{\exp(x)-\exp(-x)}{2}$. Compute sinh for $x = 2$ using exponentials. Verify that the result is indeed the hyperbolic sin using MATLAB's built-in function `sinh`.

▷▷ **16.** Verify that $\sin^2(x) + \cos^2(x) = 1$ for $x = \pi, \frac{\pi}{2}, \frac{\pi}{4}, \frac{\pi}{6}$. Use format long.

▷▷ **17.** Call the help function for the function `sind`. Use `sind` to compute the sin $87°$.

18. Write a MATLAB statement that generates the following error:
"Undefined function or method 'sni' for input arguments of type 'double'."
Hint: sni is a misspelling of the function `sin`.

19. Write a MATLAB statement that generates the following error:
"Not enough input arguments."
Hint: Input arguments refers to the input of a function (any function); for example, the input in `sin(pi/2)` is `pi/2`.

20. Write a MATLAB statement that generates the following error:
"Expression or statement is incorrect–possibly unbalanced (, {, or [."

21. If P is a logical expression, the law of noncontradiction states that P AND (NOT P) is always false. Verify this for P true and P false.

22. Let P and Q be logical expressions. De Morgan's rule states that NOT (P OR Q) = (NOT P) AND (NOT Q) and NOT (P AND Q) = (NOT P) OR (NOT Q). Generate the truth tables for each statement to show that De Morgan's rule is always true.

23. Under what conditions for P and Q is (P AND Q) OR (P AND (NOT Q)) false?

24. Construct an equivalent logical expression for OR using only AND and NOT.

25. Construct an equivalent logical expression for AND using only OR and NOT.

26. The logical operator XOR has the following truth table:
Construct an equivalent logical expression for XOR using only AND, OR, and NOT that has the same truth table (see Figure 1.3).

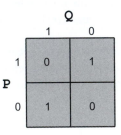

FIGURE 1.3

XOR Truth table.

27. Do the following calculation at the MATLAB command prompt. Give answers accurate to *16 digits*.
$$e^2 \sin \pi/6 + log_e(3) \cos \pi/9 - 5^3$$

28. Do the following logical and comparison operations at the MATLAB command prompt. You may assume that P and Q are logical expressions.

For $P = 1$ and $Q = 1$: Compute NOT(P) AND NOT(Q).

For $a = 10$ and $b = 25$: Compute $(a < b)$ AND $(a == b)$.

Variables and Basic Data Structures

CHAPTER OUTLINE

Motivation

Currently, technology can acquire information from the physical world at an enormous rate. For example, there are sensors that can take tens of thousands of pressure, temperature, and acceleration readings per second. To make sense of all this data and process it in a way that will help solve engineering problems requires storing information in data structures that you and MATLAB can easily work with.

Variables are used in MATLAB to store and work with data. However, data can take many forms. For example, data can be numbers, words, or have a more complicated structure. It is only natural that MATLAB would have different kinds of variables to hold different kinds of data. In this chapter, you will learn how to create and manipulate MATLAB's most common variable types.

2.1 Variables and Assignment

When programming, it is useful to be able to store information in variables. A **variable** is a string of characters and numbers associated with a piece of information. The **assignment operator**, denoted by the "=" symbol, is the operator that is used to assign values to variables in MATLAB. The line » x = 1 takes the known value, 1, and **assigns** that value to the variable with name "x." After executing this line, you will see a new variable appear in the workspace window. Until the value is changed or the variable deleted, the character x behaves like the value 1.

An Introduction to MATLAB® *Programming and Numerical Methods.* http://dx.doi.org/10.1016/B978-0-12-420228-3.00002-6

> **TRY IT!** Assign the value 2 to the variable y. Multiply y by 3 to show that it behaves like the value 2.
>
> ```
> >> y = 2
> y =
>      2
> >> y*3
> ans =
>      6
> ```

The **workspace** is an abstraction for the space in the computer's memory being utilized to store variables. For now, it is sufficient to know that the command window has its own workspace, the contents of which are made visually available in the workspace window. As a result of the previous example, you will see the variable y appear in the workspace window. You can view a list of all the variables in the command window's workspace using the function whos.

Note that the equal sign in programming is *not* the same as a truth statement in mathematics. In math, the statement $x = 2$ declares the universal truth within the given framework, x *is* 2. In programming, the statement x = 2 means a known value is being associated with a variable name, *store* 2 in x. Although it is perfectly valid to say $1 = x$ in mathematics, assignments in MATLAB always go *left*: meaning the value to the right of the equal sign is assigned to the variable on the left of the equal sign. Therefore, » 1 = x will generate an error in MATLAB. The assignment operator is always last in the order of operations relative to mathematical, logical, and comparison operators.

> **TRY IT!** The mathematical statement $x = x+1$ has no solution for any value of x. In programming, if we initialize the value of x to be 1, then the statement makes perfect sense. It means, "Add x and 1, which is 2, then assign that value to the variable x." Note that this operation overwrites the previous value stored in x.
>
> ```
> >> x = 1
> x = 1
> >> x = x + 1
> x = 2
> ```

There are some restrictions on the names variables can take. Variables can only contain alphanumeric characters (letters and numbers) as well as underscores. However, the first character of a variable name must be a letter. The maximum length of a variable name is 255 characters, which is rarely a problem. Spaces within a variable name are not permitted, and the variable names are case-sensitive (e.g., x and X will be considered different variables).

> **TIP!** Unlike in pure mathematics, variables in programming almost always represent *something* tangible. It may be the distance between two points in space or the number of rabbits in a population. Therefore, as your code becomes increasingly complicated, it is very important that your variables

> carry a name that can easily be associated with what they represent. For example, the distance between two points in space is better represented by the variable `dist` than `x`, and the number of rabbits in a population is better represented by `nRabbits` than `y`.

Note that when a variable is assigned, it has no memory of *how* it was assigned. That is, if the value of a variable, `y`, is constructed from other variables, like `x`, reassigning the value of `x` will not change the value of `y`.

EXAMPLE: What value will `y` have after the following lines of code are executed?

```
>> x = 1
x = 1
>> y = x + 1
y = 2
>> x = 2
x = 2
>> y
y = 2
```

WARNING! You can overwrite variables or functions that have been stored in MATLAB. For example, the command » `sin = 2` will store the value 2 in the variable with name `sin`. After this assignment `sin` will behave like the value 2 instead of the function *sin*. Therefore, you should always be careful not to give your variables the same name as built-in functions or values. An easy way to check if a name is already being used is by using the help function.

You can clear a variable from the workspace using the `clear` function. Typing » `clear x` will clear the variable `x` from the workspace. Typing » `clear` or » `clear all` will remove all the variables from the workspace. Typing » `clc` will clear the screen, but will not remove any of your variables.

In mathematics, variables are usually associated with unknown numbers; in programming, variables are associated with a value of a certain type. There are many data types that can be assigned to variables. A **data type** is a classification of the type of information that is being stored in a variable. The basic data types that you will utilize throughout this book are logical, double, `char`, struct, and cell. A formal description of these data types is given in the following sections.

First, we give a brief overview of matrices and their parts. A **matrix** or **array** can be viewed as a rectangular table of values, not necessarily numerical values. An **element of a matrix** is a unit of information contained in a matrix. An **index of a matrix** is an address within that array. For this book, we will be dealing exclusively with one- and two-dimensional arrays. For one-dimensional arrays, the index is a positive integer denoting the position of the element under consideration. For two-dimensional arrays, the index is a pair of positive integers that denotes the row and column of the element under consideration.

In mathematics, matrices are usually associated with tables of numbers. However, in MATLAB, every value is considered to be a matrix. Words are defined as a matrix of letters. Even a single number is considered a 1×1 matrix.

TIP! Some of the data structures that you will create and work with will be very large, having thousands or millions of entries. Since it is not useful for a human to physically view the contents of these data structures, a semicolon can be used after a variable is created to suppress the display. For example, » x = 2; will not display the resulting assignment to the screen, but the assignment to x will still be executed. You can verify this by looking in the workspace window.

TRY IT! Assign the value 2 to the variable x with a semicolon and without a semicolon after the command.

```
>> clear all
>> x = 2;
>> x = 2
x =

     2
```

TIP! Now that you know how to assign variables, it is important that you learn to *never* leave unassigned commands. An **unassigned command** is an operation that has a result, but that result is not assigned to a variable. For example, you should never use » 2 + 2. You should instead assign it to some variable » x = 2 + 2. This allows you to "hold on" to the results of previous commands and will make your interaction with MATLAB must less confusing.

2.2 Double Arrays

A **double** is a MATLAB data type used to denote numbers. Double is the most important data type you will learn about because engineers work most frequently with numbers. There are other data types used to denote numbers, but unless otherwise indicated, all numbers will be doubles for the purposes of this text unless otherwise stated. Double stands for "double-precision," which will be described in more detail in Chapter 8, on Representation of Numbers.

TRY IT! Assign the value 1 to the variable x. Verify that x is a double using the `class` function.

```
>> x = 1;
>> c = class(x)
    c = double
```

You can build up double arrays in MATLAB using square brackets, [ ]. The technical term for the function provided by brackets is called **concatenation** (noun) or **concatenate** (verb). It is common to separate columns by commas and rows by semicolons within a concatenation. You can create an empty array by placing brackets around nothing.

TRY IT! Create the following arrays:

$$x = \begin{bmatrix} 1 & 4 & 3 \end{bmatrix}$$
$$y = \begin{bmatrix} 1 & 4 & 3 \\ 9 & 2 & 7 \end{bmatrix}$$

```
>> x = [1,4,3];
>> y = [1, 4, 3; 9, 2, 7]
```
or
```
>> y = [x; 9, 2, 7]
```
or
```
>> y = [[1, 4, 3]; [9, 2, 7]]
```
or
```
>> y = [[1; 9], [4; 2], [3; 7]]
```

TRY IT! Create an empty array. Verify that it is a double using the `class` function.

```
>> x = [];
>> c = class(x)
c =
    double
```

Note that, again, MATLAB always resolves the innermost brackets first. The line » y = [[1; 9], [4; 2], [3; 7]] will concatenate [1;9] before the outer brackets. If you try to concatenate matrices that do not fit together (i.e., they do not form a rectangle), you will get an error saying that the horizontal or vertical concatenation is incorrect.

TIP! A bracket can be read as "put together." So » x = [1 4 3] can be read as "put together 1, 4, and 3 into an array, and then assign it to the variable x."

WARNING! MATLAB abuses notation by giving the semicolon two purposes. It suppresses output when at the end of a line but separates rows when making arrays, so be careful when using it.

Many times we would like to know the size or length of an array. The `size` function is called on an array M and returns a 1×2 array where the first element is the number of rows in the matrix M and the second element is the number of columns in M. Note that the output of the `size` function is also an array. The `size` function also allows you to specify the size of only one dimension using the input `size(M,dim)`. So `size(M,1)` would return the number of rows in M, and `size(M,2)` would return the number of columns in M. The `length` function is called on an array M and returns the number of elements in matrix M if M is one dimensional. If M has more than one dimension, then it returns the length of the largest dimension of M. Try not to use this feature of `length` (i.e., use `length` on one-dimensional arrays only).

TRY IT! Compute the size and length of the matrices x and y given in the previous example. Use the `size` function to obtain only the number of rows in y and only the number of columns in y.

```
>> s = size(x)
 s = 1 3
>> L = length(x)
 L = 3
>> s = size(y)
 s = 2 3
>> L = length(y)
 L = 3
>> rows = size(y,1)
 rows = 2
>> cols = size(y,2)
 cols = 3
```

Very often we would like to generate arrays that have a structure or pattern. For instance, we may wish to create the array z = [1 2 3 ... 2000]. It would be very cumbersome to type the entire description of z into the command prompt. For generating arrays that are in order and evenly spaced, it is useful to use the colon operator, "`:`".

CONSTRUCTION: Colon Operator

```
start value : increment: final value
```

Using the colon operator, z can be created with z = 1:1:2000. Since it is very common to have an increment of 1, if an increment is not specified, MATLAB will use a default value of 1. Therefore »
z = 1:2000 will have the same result as » z = 1:1:2000. Negative or noninteger increments can also be used. So z = 2:−.5:0 becomes z = [2 1.5 1 .5 0]. If the increment "misses" the last value, it will only extend until the value just before the ending value. For example, x = 1:2:8 would be [1, 3, 5, 7].

Sometimes we want to guarantee a start and end point for an array but still have evenly spaced elements. For instance, we may want an array that starts at 1, ends at 8, and has exactly 10 elements.

For this purpose you can use the function `linspace`. Unlike the functions you have worked with before that take only one input, `linspace` takes three input values separated by commas. So A = `linspace(a,b,n)` generates an array of n equally spaced elements starting from a and ending at b.

TRY IT! Use `linspace` to generate an array starting at 3, ending at 9, and containing 10 elements.

```
>> A = linspace(3,9,10)
A =
  Columns 1 through 9

    3.0000    3.6667    4.3333    5.0000    5.6667    6.3333    7.0000    7.6667    8.3
  Column 10
    9.0000
```

Another type of array that is highly structured is a matrix in which every element is the same number. For this purpose, the functions `zeros` and `ones` are useful. These functions take in a number of rows and a number of columns and return a matrix of the input dimension of zeros or ones.

TRY IT! Generate a 3 × 5 array of ones and zeros.

```
>> A = zeros(3,5)
A =
     0     0     0     0     0
     0     0     0     0     0
     0     0     0     0     0

>> B = ones(5,3)
B =
     1     1     1
     1     1     1
     1     1     1
     1     1     1
     1     1     1
```

Any array-building functions or operations can be combined to create complicated arrays. Keep in mind that MATLAB will always evaluate the innermost function or bracket set first.

TRY IT! Create the following array:

$$M = \begin{bmatrix} 0 & 0 & 0 & 1 & 1 & 1 \\ 0 & 0 & 0 & 1 & 1 & 1 \\ 1 & 2 & 3 & 4 & 5 & 6 \\ 2 & 4 & 6 & 8 & 0 & 0 \\ 8 & 7 & 2 & 5 & 9 & 0 \end{bmatrix}$$

```
>> M = [[zeros(2,3)],[ones(2,3)]; 1:6; 2:2:8,0,0; [8,7,2,5,9,0]]
M =
        0    0    0    1    1    1
        0    0    0    1    1    1
        1    2    3    4    5    6
        2    4    6    8    0    0
        8    7    2    5    9    0
```

In MATLAB, the indices of an array are denoted by parenthese attached to the variable name. If A is an array in the current workspace, we can obtain the element in row r and column c using the notation $A(r,c)$. This is referred to as **array indexing**. You can specify an array of indices to get multiple elements of an array. In other words, r and c can be arrays, and you can use array creation operations within the indexing. If you want an entire row or column, you can shorthand this operation with a colon, $:$. If you want to go to the end of an array while indexing, you can use the word end.

TRY IT! Let A = [1.5 2.5 3.5; 4.5 5.5 6.5] be in the current workspace. Find the element in the second row, third column (i.e., 6.5) using array indexing.

```
>> a = A(2,3)
a =
    6.5
```

TRY IT! Let A be as in the previous example. Retrieve the elements in the first row and the first and third columns.

```
>> a = A(1, [1,3])
a =
    1.5 3.5
```

TRY IT! Let A be as in the previous example. Retrieve all the elements in the second row of A.

```
>> a = A(2,1:3)
a =
    4.5 5.5 6.5
```

or

```
>> a = A(2,1:size(A,2))
a =
    4.5 5.5 6.5
```

or

```
>> a = A(2,1:end)
a =
    4.5 5.5 6.5
```

```
or
>> a = A(2,:)
a =
      4.5 5.5 6.5
```

Remember that MATLAB will always resolve the innermost command first. Therefore in the line `a = A(2,1:size(A,2))`, `size(A,2)` will be evaluated first, the value of which is 3 (the number of columns in A). Then `1:size(A,2)` is equivalent to `1:3` and is resolved next. Then `A(2,[1 2 3])` is resolved to `[4.5 5.5 6.5]`, and, finally, the result is assigned to the variable a.

For one-dimensional arrays, you can shorthand by only including a single index.

TRY IT! Let A = [7 3 9 2 4 5]. Write commands that retrieve the third element of A, retrieve the third, fifth, and sixth elements of A, and retrieve the third, fourth, and fifth elements of A.

```
>> a = A(3)
a =
      9
>> a = A([3, 5, 6])
a =
      9 4 5
>> a = A(3:5)
a =
      9 2 4
```

You can reassign a value of an array by using array indexing and the assignment operator. You can reassign multiple elements to a single number using array indexing on the left side. You can also reassign multiple elements of an array as long as both the number of elements being assigned and the number of elements assigned is the same. You can create an array using array indexing.

TRY IT! Let A = [1 2 3 4 5 6]. Reassign the fourth element of A to 7. Reassign the first, second, and third elements to 1. Reassign the second, third, and fourth elements of A to 9, 8, and 7.

```
>> A(4) = 7
A =
      2 3 7 5 6
>> A(1:3) = 1
A =
      1 1 1 7 5 6
>> A(2:4) = [9 8 7]
A =
      1 9 8 7 5 6
```

TRY IT! Create the matrix A = [1 2; 3 4] using array indexing.

```
>> A(1,1) = 1;
>> A(1,2) = 2;
>> A(2,1) = 3;
>> A(2,2) = 4
A =
     1 2
     3 4
```

WARNING! Although you can create an array from scratch using indexing, we do not advise it. It can confuse you and errors will be harder to find in your code later. For example, » B(2,2) = 1 will give the result B = [0 0; 0 1], which is strange because B(1,1), B(1,2), and B(2,1) were never specified.

Basic arithmetic is defined for arrays. However, there are operations between a scalar (a single number) and an array and operations between two arrays. We will start with operations between a scalar and an array. To illustrate, let a be a scalar, and M be a matrix.

M + a, M−a, M*a and M/a adds a to every element of M, subtracts a from every element of M, multiplies every element of M by a, and divides every element of M by a, respectively.

TRY IT! Let M = [1 2; 3 4]. Add and subtract 2 from M. Multiply and divide M by 2. Square every element of M. On your own, verify the reflexivity of scalar addition and multiplication: $M + a = a + M$ and $aM = Ma$.

```
>> M1 = M + 2

M1 =
     3 4
     5 6

>> M2 = M − 2
M2 =
    −1 0
     1 2

>> M3 = 2*M
M3 =
     2 4
     6 8
```

```
>> M4 = M/2
M4 =
      0.5 1
      1.5 2

>> M5 = M.^2
M5 =
      1 4
      9 16
```

Describing operations between two matrices is more complicated. Let M and P be two matrices of the same size. $M - P$ takes every element of M and subtracts the corresponding element of P. Similarly, $M - P$ subtracts every element of P from the corresponding element of M.

TRY IT! Let M = [1 2; 3 4] and P = [3 4; 5 6]. Compute M + P and M − P.

```
>> Q = M + P
Q =
      4 6
      8 10
>> Q = M − P
Q =
      −2 −2
      −2 −2
```

There are two different kinds of matrix multiplication (and division). There is **element-by-element matrix multiplication** and standard matrix multiplication. For this section, we will only show how element-by-element matrix multiplication and division work. Standard matrix multiplication will be described in Chapter 13 on Linear Algebra. MATLAB takes the * symbol to mean standard matrix multiplication. So element-by-element multiplication is denoted by .*, and it is read "dot times." For matrices M and P of the same size, M.*P takes every element of M and multiplies it by the corresponding element of P. The same is true for ./ and .^.

TRY IT! Let M = [1 2; 3 4] and P = [3 4; 5 6]. Compute M.*P, M./P, and M.^P.

```
>> Q = M.*P
Q =
      3    8
      15   24
>> Q = M./P
Q =
      .3333    .5
```

```
     .6        .6667
>> Q = M.^P
Q =
     1   16
   243 4096
```

> **WARNING!** If you accidentally forget to put a "." in front of the multiplication sign, you will likely get an error that says "inner matrix dimensions must agree." The reason for this error will become clear when matrix multiplication is described later in Chapter 12.

The **transpose** of a matrix, M, is a matrix, P, where $P(i, j) = M(j, i)$. In other words, the transpose switches the rows and the columns of M. You can transpose a matrix in MATLAB using an apostrophe, '.

> **TRY IT!** Let M = [1 2; 3 4]. Compute the transpose of M.
>
> ```
> >> P = M'
> P =
>      1 3
>      2 4
> ```

All of MATLAB's built-in arithmetic functions, such as sin, can take arrays as input arguments. The output is the function evaluated for every element of the input array. A function that takes an array as input and performs the function on it is said to be **vectorized**.

> **TRY IT!** Compute sqrt for x = [1 4 9 16].
>
> ```
> >> x = [1 4 9 16];
> >> y = sqrt(x)
> y =
>      1 2 3 4
> ```

Logical operations are only defined between a scalar and an array and between two arrays of the same size. Between a scalar and an array, the logical operation is conducted between the scalar and each element of the array. Between two arrays, the logical operation is conducted element-by-element.

> **TRY IT!** Check which elements of the array x = [1 2 4 5 9 3] are larger than 3. Check which elements in x are larger than the corresponding element in y = [0 2 3 1 2 3].

```
>> v = [1 2 4 5 9 3] > 3
v =
    0 0 1 1 1 0
>> w = [1 2 4 5 9 3] > [0 2 3 1 2 3]
w =
    1 0 1 1 1 0
```

MATLAB can index elements of an array that satisfy a logical expression.

TRY IT! Let v be the same array as in the previous example. Create a variable y that contains all the elements of x that are strictly bigger than 3. Assign all the values of x that are bigger than 3, the value 0.

```
>> y = x(x>3)
y =
    4    5    9

>> x(x>3) = 0
x =
    1    2    0    0    0    3
```

2.3 Char Arrays

Char is a data type for storing alphanumeric characters. An array of chars, usually one-dimensional, is called a **string**. Strings are assembled using apostrophes on both sides, but brackets can also be used to concatenate strings.

TRY IT! Assign the character 'S' to the variable with name s. Assign the string 'Hello World' to the variable w. Verify that s and w have the type char using the class function.

```
>> s = 'S'
s = S
>> w = 'Hello World'
w = Hello World
>> c = class(s)
c = char
>> c = class(w)
c = char
```

Note that a blank space, ' ', between 'Hello' and 'World' is also a char. Any symbol can be a char, even the ones that have been reserved for operators. Note that as a char, they do not perform the same function. Although they look the same, MATLAB interprets them completely differently.

TRY IT! Assign the `char`, `'+'`, to the variable p. Verify that p does not behave like the addition operator, +.

```
>> p = '+';
>> 1 p 2
??? 1 p 2
     |
Error: Unexpected MATLAB expression.
```

WARNING! Numbers can also be expressed as `chars`. For example, `x = '123'` means that `x` is the *string* 123 *not* the number 123. However, `chars` represent words or text and so should not have addition defined on them. MATLAB somewhat abuses this difference and allows addition between `chars` and doubles (as well as other arithmetic operations). The reason this works is that every `char` has something called a `char` code, which is how the `char` is represented in deeper levels of MATLAB. It is strongly advised that you avoid doing this.

TRY IT! Add 1 to the string 1. Add 1 to the string 123.

```
>> x = 1 + '1'
x = 50
>> y = 1 + '123'
y = 50    51    52
```

TIP! You may find yourself in a situation where you would like to use an apostrophe as a `char`. This is problematic since an apostrophe is used to denote strings. Fortunately, an apostrophe can be made by using *two* apostrophes.

TRY IT! Create the string `'don't'`.

```
>> d = 'don''t'
d = don't
```

Just as with double arrays, the size and length of a string is the number of rows and columns or the number of elements contained in it, respectively. Strings can be concatenated together, vertically and horizontally, using brackets just like doubles. However, for the purposes of this text, `chars` will always be one dimensional.

TRY IT! Create the strings s1 = 'Hello' and s2 = 'World' and use the concatenation of s1 and s2 to make the string s3 = 'Hello World'. Remember the space between the words!

```
>> s1 = 'Hello';
>> s2 = 'World';
>> s3 = [s1, ' ', s2]
s3 =
    Hello World
```

Char arrays are indexed the same as double arrays, including array indexing. An empty string can be created with two apostrophes, ' '.

TRY IT! Assign the char 'E7 is Awesome' to the variable s. Retrieve only the letter E and only the word Awesome from s using array indexing.

```
>> s = 'E7 is Awesome';
>> E = s(1)
E =
    E
>> Awesome = s(7:end)
Awesome =
    Awesome
```

TRY IT! Create an empty string. Verify that the empty string is a char.

```
>> s = '';
>> c = class(s)
c =
    char
```

A very useful function in MATLAB is sprintf. The sprintf function writes new data to a preformatted string.

TRY IT! Use the sprintf function to make the strings s1 = 'My name is Timmy' and s2 = 'My name is Alex'.

```
>> s1 = sprintf('My name is %s', 'Timmy')
s1 =
    My name is Timmy

>> s2 = sprintf('My name is %s', 'Alex')
s2 =
    My name is Alex
```

> **WHAT IS HAPPENING?** In the previous example, the first input to `sprintf` is a string of the desired format. In each case, the place where the name should be can be different. Therefore, a `%s` is placed wherever the name should be (s stands for string in this case).

You can make longer formatted strings as well with more than one placeholder. You can use `sprintf` for inserting strings `%s`, integers `%d`, or more generic numbers `%f` and `%g`. You can also control the number of digits inserted into the formatted string, but we will leave that to you to explore on your own.

> **TRY IT!** Use the `sprintf` function to make the string `s = 'This is E7 and there are 423 students in the class'`.
>
> ```
> >> s = sprintf('This is %s and there are %d students in the class', 'E7', 423)
> s =
>     This is E7 and there are 423 students in the class
> ```

> **TIP!** There are functions that are said to take **sprintf type inputs**. This means that the function can take in formatted strings in the same way as `sprintf`.

2.4 Struct Arrays

A **struct** is a data type that is useful when each element of an array is defined by several properties. For example, you may want each element of an array to represent a person defined by a name and a personal ID number. However, each person may be difficult to define by only one `char` or `double`. In a `struct` array, the properties of each element are defined by its **fields**. The data type contained in a field can be of any data type. The `struct` array name and the field are separated by a period. As with `double` and `char` arrays, the index or indices are specified in parentheses next to the array name. There are ways of making a structure array all at once. However, it is more straightforward to create them using array indexing, which is not recommended for `double` or `char` arrays.

> **TRY IT!** Let a student be defined by a name (`char`), personal ID number (`double`), and an array of grades (`double`). Populate a `struct` array `student` with field names, ID, and grades. Verify that `student` has type `struct` using the `class` function.
>
> ```
> >> student(1).name = 'Abby';
> >> student(1).ID = 18309481;
> >> student(1).grades = [95 84 91 99 100];
> >> student(2).name = 'Bobby';
> >> student(2).ID = 17283097;
> ```

```
>> student(2).grades = [89 99 75 100 95];
>> student(3).name = 'Christopher';
>> student(3).ID = 17209384;
>> student(3).grades = [74 79 86 92 82];
>> student
student =
1x3 struct array with fields:
    name
    ID
    grades

>> c = class(student)
c = struct
```

TIP! Although it is possible to populate all the student names first, then all the IDs, and so on, it is advisable to finish populating the fields of a single element first before moving on.

Struct arrays can be two or more dimensions but for the purposes of this course, structs will always be one dimensional. Since structs are arrays, they can be concatenated, but only if the two struct arrays have the same fields (and are of compatible size). Also, addition and other arithmetic operations are not allowed between structs or between structs and numbers.

TRY IT! Populate a new struct called newStudent with the same fields as the student struct in the previous example. Concatenate newStudent to the student struct.

```
>> newStudent.name = 'Denise';
>> newStudent.ID = 20912834';
>> newStudent.grades = [NaN 82 99 75 95];
>> student = [student, newStudent];

student =
1x4 struct array with fields:
    name
    ID
    grades
```

Struct arrays are indexed using parentheses between the struct name the period separating the field name. You can use array indexing the same as with double and char arrays.

TRY IT! Retrieve the second element of the student struct in the previous example. Retrieve the first, second, and third element of student as well.

```
>> s = student(2)
s =
      name: 'Bobby'
        ID: 17283097
    grades: [89 99 75 100 95]
>> s = student(1:3)
1x3 struct array with fields:
    name
    ID
    grades
```

The information contained within the field of an element of a struct array can be retrieved by indexing the desired element, then placing a dot, and then typing the name of the field. For example, `student(2).grades` will return information contained in the grades field of the second element of student. The values contained in the fields of structs arrays retain their original data type; therefore, they behave exactly as that data type. For example, `student(2).grades` is the double `[89 99 75 100 95]`. Therefore, we can use array indexing to get specific elements of the second student's grades.

TRY IT! Retrieve the grades for the second student in the struct array student in the previous example. Verify that the grades are `double` using the `class` function. Retrieve only the last element of second student's grades using array indexing on the grades field of student.

```
>> grades = student(2).grades
grades =
    89 99 75 100 95
>> g = class(grades)
g =
    double
>> final = grades(end)
final =
    95
```

As stated before, the fields of a struct array can have any data type, including a `struct` or `cell` (defined in the next section). Just remember the field of a struct behaves exactly like that data type.

TRY IT! Populate a `struct` called `myActivities` with fields `sports`, `afterSchool`, and `clubs`. Assign a field called activities to the first element of the previous student `struct`. Verify that the activities field of student is a `struct` using the `class` function. Assign the data contained in the `club` field in the activities of the student `struct` to the variable `club`.

```
>> myActivities.sports = 'Basketball';
>> myActivities.afterSchool = '';
>> myActivities.clubs = 'Students for Programming';
```

```
>> student(1).activities = myActivities
student =
1x4 struct array with fields:
    name
    ID
    grades
    activities

>> a = class(student(1).activities)
a =
    struct

>> club = student(1).activities.clubs
club =
    Students for Programming
```

WARNING! If a `struct` has more than one element and you omit the indexing when calling a field, MATLAB will treat the command as if you called the field for each element individually. For example, typing `student.name` will have the following result:

```
>>student.name
ans =
Abby
ans =
Bobby
ans =
Christopher
ans =
Denise
```

Avoid doing this when coding because you can get unexpected results.

2.5 **Cell Arrays**

A **cell** is a data type for unstructured information. In a `cell` array, each element can have any data type, including another `cell`. However, elements of `cell` arrays are indexed using braces, {}, rather than parentheses. They are also concatenated using braces. Alternatively, you can create a `cell array` using the cell function, then populate the elements one by one.

TRY IT! Create a 1 × 3 `cell` array where the first element is the string 'E7', the second element is the `double` 2011, and the third element is the `struct` array student created in a previous section on `struct` arrays. Create the `cell` array using concatenation. Verify that the created `cell` array has type `cell` using the function `class`. Verify that each of the elements has type `char`, `double`, and `struct`, respectively. Repeat this example by first calling the `cell` function, then populating the elements one by one.

```
>> C = {'E7', 2011, student}
>> c = class(C)
c = cell
>> s = class(C{1})
s = char
>> d = class(C{2})
d = double
>> S = class(C{3})
S = struct

>> C = cell(1,3);
>> C{1} = 'E7';
>> C{2} = 2011;
>> C{3} = student;
>> c = class(C)
c = cell
>> s = class(C{1})
s = char
>> d = class(C{2})
d = double
>> S = class(C{3})
S = struct
```

A common mistake to make when using `cell` arrays is to index the `cell` array with parentheses rather than braces. If you do this, you will get a `cell` array containing the contents of the indexed elements.

TRY IT! Index the `cell` array from the previous example using parentheses rather than braces. Verify that contents are the cell arrays rather than the expected data types.

```
>> C = {'E7', 2011, student}
>> s = class(C(1))
s = cell
>> d = class(C(2))
d = cell
>> S = class(C(3))
S = cell
```

> **WARNING!** Although using parentheses rather than braces can be useful, we advise against doing so while you are learning to program in MATLAB.

> **TIP!** Usually data will have some kind of structure to it. As a result, it is usually better to have data stored in a struct rather than a cell.

There are many quirky behaviors associated with `cell` arrays. For this book, we will use only the rudimentary creation and manipulation operations discussed in this section.

Summary

1. Storing, retrieving, and manipulating information and data is important in any engineering field.
2. Variables are an important tool for handling data values. In MATLAB all variables are arrays.
3. There are four basic data types for storing information in MATLAB: `double` (numbers), `char` (words), `struct` (structured information), and `cell` (unstructured information).

Vocabulary

array	data type	struct
array indexing	double	transpose
assign	element of a matrix	unassigned command
assignment operator	element-by-element matrix multiplication	variable
cell	field	vectorized
char	index of a matrix	workspace
concatenate	matrix	
concatenation	sprintf type input	
data type	string	

Functions and Operators

.	fliplr	rand
:	getfield	randn
;	i	rmfield
=	Inf	setfield
[]	isfield	size
{}	length	sprintf

```
cell              linspace          strcmp
clc               lower             strcmpi
clear             NaN               upper
clear all         num2str           whos
fieldnames        ones              zeros
```

Problems

⊠ **1.** Assign the value 2 to the variable x and the value 3 to the variable y. Clear just the variable x. Then clear all the variables using `clear all`. Clear the screen using `clc`.

● **2.** Write a line of code that generates the following error:

```
??? Undefined function or variable 'x.'
```

● **3.** Write a line of code that generates the following error:

```
The expression to the left of the equals sign is not a valid
target for an assignment.
```

⊠ **4.** Let x = 10 and y = 3 be defined in the workspace. Write a line of code that will make each of the following assignments.

$$u = x + y$$
$$v = xy$$
$$w = x/y$$
$$z = \sin x$$
$$r = 8 \sin x$$
$$s = 5 \sin xy$$
$$p = x^y.$$

⊠ **5.** Let x = [1 4 3 2 9 4] and y = [2 3 4 1 2 3]. Compute the assignments from Problem 4. Remember to use array operations!

✎ **6.** Recall that `linspace(a,b,n)` generates an array of n evenly spaced numbers starting at a and ending at b. Given a, b, and n, write a statement in terms of a, b, and n using the colon operator that produces the same array as `linspace(a,b,n)`.

⊠ **7.** Create the following matrix in a single assignment. Try to use as few numbers as possible.

$$M = \begin{bmatrix} 0 & 0 & 0 & 0 & 0 & 1 \\ 0 & 0 & 0 & 0 & 0 & 1 \\ 1 & 2 & 3 & 4 & 5 & 1 \\ 0 & 2 & 4 & 6 & 8 & 1 \\ 8 & 7 & 2 & 5 & 9 & 1 \end{bmatrix}$$

⊠ **8.** For the matrix, M, in the previous problem, use the `sum` function to compute the sum of each of the *rows* in M. Hint: Use the `sum` function.

⊠ **9.** Use the `rand` function to generate a uniformly distributed array of 1000 numbers between 0 and 1. The mean of randomly distributed numbers between 0 and 1 should be 0.5. Use the `mean` function to find the mean of the array created. Verify that it is close to 0.5.

⊠ **10.** Assign the string '123' to the variable S. Use the function `str2num` to change S into a double. Assign the output of `str2num` to the variable N. Verify that S is a `char` and N is a `double` using the `class` function.

⊠ **11.** Assign the string `'HELLO'` to the variable s1 and the string `'hello'` to the variable s2. Use the `strcmp` function to compare s1 and s2 to show that they are not equal. Use the `strcmp` function to show that s1 and s2 are equal if the `lower` function is used on s1. Use the `strcmp` function to show that s1 and s2 are equal if the `upper` function is used on s2.

⊠ **12.** Use the `sprintf` function to generate the following strings.
```
The word 'Engineering' has 11 letters.
The word 'Book' has 4 letters.
The word 'MATLAB' has 6 letters.
```

⊠ **13.** Let x = `0:10` and y = `10:-1:0`. Use the equality comparison operator and the `all` function to show that all the elements of x equal all the elements of y after the `fliplr` function is used on y. Try performing the same operation using the `isequal` function.

14. Let x = `linspace(1,10,100)`. What will be the output of » `y = class(size(x))` and »z = `size(class(x))`? What about » `v = class(class(size(x)))` and » `w = size(size(class(x)))`?

15. Write lines of code that generate each of the following array-related errors:
```
??? Error: Unbalanced or unexpected parenthesis or bracket.
??? Attempted to access A(-1); index must be a positive integer
or logical.
??? Error using ==> horzcat CAT arguments dimensions are not
consistent.
??? Error using ==> vertcat CAT arguments dimensions are not
consistent.
??? In an assignment A(I) = B, the number of elements in B and
I must be the same.
??? Error using ==> plus Matrix dimensions must agree.
??? Attempted to access A(4); index out of bounds because
numel(A)=3.
```

16. Write lines of code that generate each of the following struct-related errors:
```
??? Error using ==> horzcat CAT arguments are not consistent in
structure field names.
??? Undefined function or method 'plus' for input arguments of
type 'struct'.
```

17. Let cars be a struct where each element represents a different car. List some fields that cars should have given that you are (a) a car salesman and (b) an engineer.

18. Create a `struct` array called Class with fields title (`string`), semester (`string`), and enrollment (`double`). Populate the Class struct with information from three of your favorite courses.

Example first element:

```
Class(1).title = 'E7';
Class(1).semester = 'Spring 2011';
Class(1).enrollment = 405;
```

For the example first element given, use concatenation of the field values in Class to generate the string `'E7: Spring 2011'`.

19. Create a `cell` array called Class where each row contains data about a class. The first element of the row should be the title of the course (string), the second element should be the semester of the course (string), and the third element of the row should be the enrollment of the course (double). Populate the `Class` cell with information from three pieces of information from your favorite course.

Example first row:

```
Class{1,1} = 'E7';
Class{1,2} = 'Spring 2011';
Class{1,3} = 405;
```

For the example first row given, use concatenation of the elements in Class to generate the string `'E7: Spring 2011'`.

20. Write a command that assigns the string `Hello World` to the variable S.

21. Let $S1$ and $S2$ be strings in the current workspace. Write a command that uses the `strcmp` function to check if they are the same.

22. Write a command that will return the size of the array S created in the previous problem.

23. Write a command that will return the length of the `person` struct.

24. Write a command that will access the contents of the second element of the `person` struct.

25. Write a command that uses the `class` function to determine the data type of the contents in the `name` field of the second element in the `person` struct.

26. Write a command that creates an empty 1×3 cell array. Use the `cell` function.

27. Write a command that creates a 2×2 cell array where the upper left element contains the string `dog`, the upper right element contains the string `cat`, the lower left element contains the double 10, and the lower right element contains the array `[23 5 4 93]`.

28. Write a command that assigns the value $\pi/4$ to the variable x.

29. Write a command that assigns a 1×100 array of zeros to the variable z.

30. Write a command that assigns the array `A=[1 3 5 . . . 19 21]`.

31. Write a single command that will access the third, fourth, and seventh elements of `A`.

32. Write a command that assigns an array starting at 17, ending at 23, and containing 101 elements to the variable M.

33. Write a command that will give the last element of M.

34. Let A and B be $1 \times n$ arrays in the current workspace. Write a command that will horizontally concatenate them and a command that will vertically concatenate them.

35. Let A be a double array in the current workspace. Write a command that will return an array consisting of all the elements of A that are zero. Hint: Use the find function and logical operations.

36. Write the commands that will clear all the variables in the workspace and clear the screen.

Functions

Motivation

Programming often requires repeating a set of tasks over and over again. For example, the `sin` function in MATLAB is a set of tasks (i.e., mathematical operations) that computes an approximation for sin (x). Rather than having to retype or copy these instructions every time you want to use the sin function, it is useful to store this sequence of instruction as a function that you can call over and over again.

Writing your own functions is the focus of this chapter, and it is the single most powerful use of computer programming. By the end of this chapter, you should be able to declare, write, store, and call your own functions.

3.1 Function Basics

In programming, a **function** is a sequence of instructions that performs a specific task. A function can have **input arguments**, which are made available to it by the **user**, the entity calling the function. Functions also have **output arguments**, which are the results of the function that the user expects to receive once the function has completed its task. For example, the MATLAB function `sin` has one input argument, an angle in radians, and one output argument, an approximation to the sin function computed at the input angle (rounded to 16 digits). The sequence of instructions to compute this approximation constitute the **body of the function**, which until this point has not been shown.

A function can be specified in several ways. A function can be specified mathematically; for example, $f : x \rightarrow \sin x$, which means f is the function that takes x and returns $\sin x$. A function can be

An Introduction to MATLAB® Programming and Numerical Methods. http://dx.doi.org/10.1016/B978-0-12-420228-3.00003-8

defined by its **type definition**, a list of its input and output arguments by data type; for example, sin: double → double, which means `sin` is a function that takes doubles as input arguments and returns doubles as output arguments.

In this book, we will define functions in terms of its **function header**. A function header is the way a function's type definition is given to MATLAB. The function header is a list of the function's output arguments, surrounded by brackets, followed by an equal sign, the function's name, and then the function's input arguments, surrounded by parentheses. For example, the `sin` function's header looks like this: `[y] = sin(x)`.

For more complicated functions, the type definition and function header will usually be followed by a brief description of what the function should do (i.e., the relationship between the input and output arguments).

EXAMPLE: What is the type definition and function header of the `linspace` function? What is the type definition and function header of the `strcmp` function?

`linspace`: double × double × double → double.
`[A] = linspace(a,b,n)`.
`strcmp`: char × char → logical.
`[out] = strcmp(s1,s2)`.

To program your own functions, you will need to use a new part of the MATLAB environment called the **editor**. The editor allows you to build, edit, and save your functions. You can open the editor by clicking on the "new m-file" button, ▯, in the upper left-hand corner of the MATLAB environment. Figure 3.1 shows the MATLAB Editor.

We will start by walking through a construction of a very simple function defined by `[out] = myAdder(a,b,c)`, where out is the sum of `a`, `b`, and `c`.

The first line of a new function should always be the word "function," followed by its function header.

CONSTRUCTION: Function header (first line of function).

`function [outputs] = function_name(inputs)`

The first line of `myAdder` is then:

`function [out] = myAdder(a,b,c)`

You may notice that the word function turns blue. This word turns blue because "function" is a **keyword**. Keywords are words that MATLAB has reserved to carry a specific meaning. In this case, the word function is reserved to denote the start of a function. Other keywords will be defined in later chapters. Keywords may not be assigned as variable or function names. For example, `>> function = 2` would produce an error.

FIGURE 3.1

The MATLAB editor.

Next, you must write the function body. This is the sequence of instructions that will produce the desired outputs based on the given inputs. If there are multiple lines within a function, MATLAB will execute them in order. Since our function is very simple, the body will require only a single line.

```
function [out] = myAdder(a,b,c)

out = a + b + c;

end
```

Note the semicolon at the end of the line `out = a + b + c`. This is very important. In most programming languages, if you want something printed to the screen (i.e., to show up on the screen) when a function is run, you have to specifically instruct the computer to do so. However, MATLAB abuses this convention and prints out *any* unsuppressed line of code to the screen. As will be demonstrated later, this can make things very confusing so you must take care to suppress all the instructions within a function. Use the `display` function or functions like it when you want a function to print to the screen.

WARNING! Always suppress code within a function by using a semicolon at the end of all assignment statements.

Second, notice that the word "end" is placed at the end of the function. The word "end" turns blue just like the keyword "function." This is because end is also a keyword. In this case, it denotes the end of a function. Later you will see that it ends many other things as well. This function will still work if end is not placed there, but it will cause problems later on. Therefore, for our purpose, always end a function with the keyword end.

WARNING! Always end a function with an end statement.

Before trying your new function, we now introduce some good coding practice. A **comment** is a line within a function that is *not* read as code. That is, MATLAB will skip over it when running your function. You can denote a comment by placing a % symbol at the beginning of a line. You will notice that any characters in that line turn green. MATLAB will not execute any code that is green. When your code becomes longer and more complicated, comments help you and those reading your code to navigate through it and understand what you are trying to do. Getting in the habit of commenting frequently will help prevent you from making coding mistakes, understand where your code is going when you write it, and find errors when you make mistakes. It is also customary to put a description of the function as well as its type definition, author, and creation date in a comment under the function header. We highly recommend that you comment heavily in your own code. We add comments to myAdder.

```
function [out] = myAdder(a,b,c)
% [out] = myAdder(a,b,c)
% out is the sum of a,b, and c.
% author
% date

% assign output
out = a + b + c;

end
```

TIP! For PC users, you can "comment out" large blocks of code by highlighting a block of code with the mouse and then pressing ctrl+r. For MAC users, you can produce the same effect with command+/. You can uncomment the code by pressing ctrl+t for PCs and command + t for MACs. Commenting large blocks of code is useful when you want to try a different solution to a problem without risk of losing information from your previous attempt.

TIP! Build good coding practices by giving variables and functions descriptive names, commenting often, and avoiding extraneous lines of code.

For contrast, consider the following function that performs the same task as myAdder but is constructed poorly. As you can see, it is extremely difficult to see what is going on and the intention of the author.

EXAMPLE: Poor representation of myAdder.

```
function [x] = myAdderPoor(y,s2,m)

z = y + s2;
z = z + m;

x = z;
end
```

Now the function is complete according to the given specifications and is well commented. To save your function, click the save button, 🖫 , in the upper left-hand corner of the editor. You can also save your function by pressing Alt → f → s or ctrl+s. When prompted to name the file, the filename *must* have the same name as the function name. The file type should be .m, an **m-file**, which is the standard file type for MATLAB functions. Save this function as myAdder.m.

Functions must conform to a naming scheme similar to variables. They can only contain alphanumeric characters and underscores, and the first character must be a letter. The name of the function should be less than 255 characters long. Once your function is saved in the current working directory, it behaves exactly as one of MATLAB's built-in functions and can be called from the command prompt or by other functions.

TIP! It is good programming practice to save often while you are writing your function. In fact many programmers report saving using the shortcut ctrl+s every time they stop typing!

TRY IT! Use your function myAdder at the command prompt to compute the sum of a few numbers. Verify that the result is correct. Try calling the help function on myAdder.

```
>> d = myAdder(1,2,3)
d = 6
>> d = myAdder(4,5,6)
d = 15
>> help myAdder
   [out] = myAdder(a,b,c)
   out is the sum of a,b, and c.
   author
   date
```

WHAT IS HAPPENING? First recall that the assignment operator works from right to left. This means that myAdder(1,2,3) is resolved before the assignment to d.

1. MATLAB finds the function myAdder.
2. myAdder takes the first input argument value 1 and assigns it to the variable with name a (first variable name in input argument list).
3. myAdder takes the second input argument value 2 and assigns it to the variable with name b (second variable name in input argument list).
4. myAdder takes the third input argument value 3 and assigns it to the variable with name c (third variable name in input argument list).
5. myAdder computes the sum of a, b, and c, which is $1 + 2 + 3 = 6$.
6. myAdder assigns the value 6 to the variable out.
7. myAdder reaches the end of the function, identified by the keyword end.
8. myAdder verifies that a variable with name out (first variable name in the output argument list) has been created.
9. myAdder outputs the value contained in the output variable out, which is 6.
10. myAdder(1,2,3) is equivalent to the value 6, and this value is assigned to the variable with name d.

MATLAB gives the user tremendous freedom to assign variables to different data types. For example, it is possible to give the variable x a struct value or a double value. In other programming languages this is not always the case, you must declare at the beginning of a session whether x will be a struct or a double, and then you're stuck with it. This can be both a benefit and a drawback (more on this in Chapter 9). For instance, myAdder was built assuming that the input arguments were doubles. However, the user may accidently input a struct or cell into myAdder, which is not correct. If you try to input a nondouble input argument into myAdder, MATLAB will continue to execute the function until something goes wrong or until the function ends.

TRY IT! Use the string '1' as one of the input arguments to myAdder. Also use a struct as one of the input arguments to myAdder.

```
>> d = myAdder('1', 2, 3)
d =
    54

>> x.y = 2;
>> d = myAdder(1,2,x)
??? Undefined function or method 'plus' for input arguments of type 'struct'.

Error in ==> myAdder at 3
out = a + b + c;
```

> **TIP!** Remember to read the errors that MATLAB gives you. They usually tell you exactly where the problem was. In this case, the error says `Error in ==> myAdder at 3`, meaning there was an error in `myAdder` on the third line. The reason there was an error is because the variable 'c' was assigned a struct and then added to a double, which is undefined.

At this point, you do not have any control over what the user assigns your function as input arguments and whether they correspond to what you intended those input arguments to be. So for the moment, write your functions assuming that they will be used correctly. You can help yourself and other users use your function correctly by commenting your code well.

You can **compose** functions by assigning function calls as the input to other functions. In the order of operations, MATLAB will execute the innermost function call first. You can also assign mathematical expressions as the input to functions. In this case, MATLAB will execute the mathematical expressions first.

TRY IT! Use the function `myAdder` to compute the sum of $\sin(\pi)$, $\cos(\pi)$, and $\tan(\pi)$. Use mathematical expressions as the input to `myAdder` and verify that it performs the operations correctly.

```
>> d = myAdder(sin(pi), cos(pi), tan(pi))
d =
    -1

>> d = myAdder(5+2, 3*4, 12/6)
d =
    21

>> d = (5 + 2) + 3*4 + 12/6
d =
    21
```

MATLAB functions can have multiple output arguments. The example demonstrates how to write and call a function that has multiple output arguments and how to make assignments to all of its outputs. When calling a function with multiple output arguments, you can place a list of variables you want assigned inside brackets separated by commas. Consider the following function (note that it has multiple output arguments):

```
function [A,B] = myTrigSum(a,b)
% [A,B] = myTrigSum(a,b)
% A = sin(a) + cos(b)
% B = sin(b) + cos(a)
% author
% date
```

```
A = sin(a) + cos(b);
B = sin(b) + cos(a);
end
```

TRY IT! Compute the function `myTrigSum` for a = 2 and b = 3. Assign the first output argument to the variable C and the second output argument to the variable D.

```
>> [C,D] = myTrigSum(2,3)
C =
   -0.0807
D =
   -0.2750
```

If you make less assignments than there are output variables, MATLAB will make only the first assignments and the rest will be dropped. Try not to do this unless you specifically know that you want the output value to be ignored.

TRY IT! Compute the function `myTrigSum` for a = 2 and b = 3. Assign the first output argument to the variable C and drop the second output argument.

```
>> C = myTrigSum(2,3)
C =
   -0.0807
```

When writing functions, you may forget to assign one of the outputs if your function is complicated. If this is the case, MATLAB will stop, and you will get an error. Consider the following erroneous code where the output variable B is unassigned.

```
function [A,B] = myTrigSum(a,b)
% [A,B] = myTrigSum(a,b)
% A = sin(a) + cos(b)
% B = sin(b) + cos(a)
% author
% date

A = sin(a) + cos(b);
end
```

TRY IT! Run the previous erroneous code for the same inputs as in the previous example. Take note of the reported error.

```
>> [C,D] = myTrigSum(2,3)
Error in ==> myTrigSum at 8
A = sin(a) + cos(b);

??? Output argument "B" (and maybe others) not assigned during call to
"C:\Users\myTrigSum.m>myTrigSum".
```

Another useful keyword is `return`. When MATLAB sees a `return` statement, it *immediately* terminates the function as if it had executed the end statement of the function. If any of the output arguments have not been assigned when a `return` statement is made, you will get an error.

3.2 **Function Workspace**

Chapter 2 introduced the idea of the workspace where variables created at the command prompt are stored. A function also has a workspace. A **function workspace** is a space in computer memory that is reserved for variables created within that function. This workspace is not shared with the command window's workspace. Therefore, a variable with a given name can be assigned within a function without changing a variable with the same name outside of the function. A function workspace is opened every time a function is used.

TRY IT! What will the value of `out` be after the following lines of code are executed? Note that it is not 6, which is the value `out` is assigned inside of `myAdder`.

```
>> out = 1;
>> d = myAdder(1,2,3);
>> out
out =
     1
```

In `myAdder`, the variable `out` is a **local variable**. That is, it is only defined in the function workspace of `myAdder`. Therefore, it cannot affect variables in workspaces outside of the function, and actions taken in workspaces outside the function cannot affect it, even if they have the same name. So in the previous example, there is a variable, `out`, defined in the command window workspace. When `myAdder` is called on the next line, MATLAB opens a new workspace for that function's variables. One of the variables created in this workspace is another variable, `out`. However, since they have different workspaces, the assignment to `out` inside `myAdder` does not change the value assigned to `out` in the command window workspace.

This is one of the reasons it is dangerous to keep lines of code unsuppressed in your functions. We modify the function `myAdder` so that it does not have a semicolon after the line `out = a + b + c`, and has the extraneous command assigning the value 2 to the variable `y` (also unsuppressed).

```
function [out] = myAdder(a,b,c)
% [out] = myAdder(a,b,c)
% MODIFIED myAdder
% out is the sum of a,b, and c.
% author
% date

out = a + b + c

y = 2

end
```

> **TIP!** With the modified myAdder, what is the value of out after the following code is executed?
>
> ```
> >> out = 1;
> >> d = myAdder(1,2,3);
> out = 6
> y = 2
> >> out
> out = 1
> >> y
> undefined function or variable y
> ```

In the previous example, the value of out was displayed when the line out = a + b + c was executed inside myAdder. This means that the out variable being displayed is the out from myAdder's workspace, not from the command window workspace where the out variable has value 1. Therefore, the assignment to out inside the function did not affect the out variable on the outside. Likewise, the variable y was created inside of the function workspace and is not an output variable. Therefore it does not exist in the command prompt workspace. So for a final warning, be sure to suppress instructions inside a function.

> **WARNING!** If you intend to keep the function myAdder stored on your own computer, make sure to return it to the correct, original configuration.

Why have separate function workspaces rather than a single workspace? Although it may seem like a lot of trouble for MATLAB to separate workspaces, it is very efficient for large projects consisting of many functions working together. If one programmer is responsible for making MATLAB's sin function, and another for making MATLAB's cos function, we would not want each programmer to have to worry about what variable names the other is using. We want them to be able to work independently and be confident that their own work did not interfere with the other's and vice versa. Therefore, separate workspaces protect a function from outside influence. The only things from outside the function's workspace that can affect what happens inside a function are the input arguments, and the only things that can escape to the outside world from a function's workspace when the function terminates are the output arguments.

The next examples are designed to be exercises in function workspace. They are intentionally very confusing, but if you can untangle them, then you probably understand function workspace. Focus on *exactly* what MATLAB is doing, in the order that MATLAB does it.

EXAMPLE: Consider the following function:

```
function [x,y] = myWorkspaceTest(a,b)
% [x,y] = myWorkspaceTest(a,b)
% Does miscellaneous arithmetic operations on a and b
% author
% date

x = a + b;
y = x * b;

z = a + b;

end
```

TRY IT! What will the value of a, b, x, y, and z be after the following code is run?

```
>> clear all
>> a = 2;
>> b = 3;
>> z = 1;
>> [y,x] = myWorkspaceTest(b,a)
y = 5
x = 10
>> z
z = 1
```

TRY IT! What will the value of a, b, x, y, and z be after the following code is run?

```
>> x = 5;
>> y = 3;
>> [b,a] = myWorkspaceTest(x,y)
b = 8
a = 24
>> z
??? Undefined function or variable 'z'.
```

3.3 MATLAB®'s Path

When a function is called at the command prompt or from within a function, MATLAB must look for the m-file that tells it how to execute that function. The way MATLAB looks for this function is through the MATLAB **path**, which is the order in which MATLAB searches for a function when it is called. When MATLAB is asked to execute a function, it first looks for that function in the working directory or as a subfunction (described in the next section). If it is not in the working directory or a subfunction, MATLAB will look in folders along the MATLAB path until it finds the appropriate m-file to execute. If MATLAB gets to the end of the path without finding the requested function, it returns an error saying that the requested function cannot be found.

> **TRY IT!** Try to call the nonexistent function abcdefg for input value, 4, and assign the output to the variable, a.
>
> ```
> >> a = abcdefg(4)
> ??? Undefined function or method 'abcdefg' for input arguments of type 'double'.
> ```

If there are two functions with the same name in MATLAB's path, MATLAB will execute the function first in the MATLAB path. Writing a function with the same name as another function and placing it higher in the MATLAB path is called **overloading** a function. Overloading also refers to

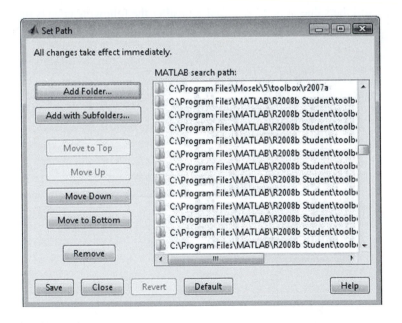

FIGURE 3.2

The MATLAB path editor.

giving variables the same name as functions you have written. Overloading is useful when you want to do a task differently than normal during a particular project. For example, you may overload the `sin` function with your own function that computes sin to 100 digits of accuracy. However, it is easy to overload functions that you do not intend to, such as naming a function or variable `gradient`, which is in fact a MATLAB function. Therefore, it is important to give your function a name that will most likely be unique. In this book, we add the word "my" to the beginning of every function name to avoid overloading any MATLAB functions.

You can view the MATLAB path by clicking File → Set Path. The path is shown in Figure 3.2. You can modify the path under set MATLAB path. However, for our purpose we will never do this.

3.4 **Subfunctions**

Once you have created and saved a new function, it behaves just like any other MATLAB built-in function. You can call the function from the command prompt, and any other function can call on the function as well. A **subfunction** is a function that is defined in the same m-file as its **parent function**. Only the parent function is able to call the subfunction. However, the subfunction retains a separate workspace from its parent function. A subfunction is declared after the end statement of its parent function, and it must have an end statement at the end of its own definition.

TRY IT! Consider the following function and subfunction saved in a single file called `myDistXYZ.m`:

```
function [D] = myDistXYZ(x,y,z)
% [D] = myDistXYZ(x,y,z)
% x,y,z are 2D coordinates contained in 1x2 arrays.
% out(1) is the distance between x and y
% out(2) is the distance between x and z
% out(3) is the distance between y and z
% author
% date

D(1) = myDist(x,y);
D(2) = myDist(x,z);
D(3) = myDist(y,z);

end % end myDistXYZ

function [D] = myDist(x,y)
% subfunction for myDistXYZ
% D is the distance between x and y, computed using the distance formula
```

```
% distance formula computed for input 2D coordinates a and b.
D = sqrt((x(1)-y(1))^2 + (x(2)-y(2))^2);

end % end myDist
```

Notice that the variables D, x, and y appear in both myDistXYZ and myDist. This is permissible because a subfunction has a separate workspace from its parent function. Subfunctions are useful when a task must be performed many times within the function but not outside the function. In this way, subfunctions help the parent function perform its task without cluttering the working directory with extra m-files.

TRY IT! Call the function myDistXYZ for x = [0 0], y = [0 1], z = [1 1]. Try to call the subfunction myDist from the command prompt.

```
>> D = myDistXYZ([0 0], [0 1], [1 1])
D =
     1 1.4142 1
>> D = myDist([0 0], [0 1])
??? Undefined function or variable 'myDist'.
```

Following is the code repeated without using subfunctions. Notice how much busier and cluttered the function looks and how much more difficult it is to understand what is going on. Also note that this version is much more prone to mistakes because you have three chances to mistype the distance formula. It is worth noting that this function could be written more compactly using vector operations. We leave this as an exercise.

```
function [D] = myDistXYZ(x,y,z)
% [D] = myDistXYZ(x,y,z)
% x,y,z are 2D coordinates contained in 1x2 arrays.
% out(1) is the distance between x and y
% out(2) is the distance between x and z
% out(3) is the distance between y and z
% author
% date

out(1) = sqrt((x(1)-y(1))^2 + (x(2)-y(2))^2);
out(2) = sqrt((x(1)-z(1))^2 + (x(2)-z(2))^2);
out(3) = sqrt((y(1)-z(1))^2 + (y(2)-z(2))^2);

end % end myDistXYZ
```

Note that MATLAB will look through the list of subfunctions before it goes to the MATLAB path. Therefore, if a subfunction overloads another function in the MATLAB path, the subfunction will take precedence over that function when called by its parent function.

3.5 Function Handles

Up until now, you have assigned various data structures to variable names. Being able to assign a data structure to a variable allows us to pass information to functions and get information back from them in a neat and orderly way. Sometimes it is useful to be able to pass a function as a variable to another function. In other words, the input to some functions may be other functions. To accomplish this, we need **function handles**. Function handles are variables that have been assigned functions as their value. A function handle is created by placing an @ symbol in front of a function in the current path and then using the assignment operator.

TRY IT! Assign the function exp to the variable F. Verify that F has type function handle using the class function.

```
>> F = @exp;
>> c = class(F)
c =
    function_handle
```

In the previous example, F is now equivalent to the exp function. Just like x = 1 means that x and 1 are interchangeable, F and the exp function are now interchangeable.

TRY IT! Compute e^2 using the function handle F. Verify that this is correct using the exp function.

```
>> y = F(2)
y = 7.3891
>> y = exp(2)
y = 7.3891
```

TRY IT! Program a function called myFunPlusOne that takes a function handle, F, and a double x as input arguments. myFunPlusOne should return F evaluated at x, and the result added to the value 1. Verify that it works for various functions and values of x.

```
function [y] = myFunPlusOne(F,x)
% [y] = myFunPlusOne(F,x)
% F is a function handle and y = F(x) + 1
```

```
y = F(x) + 1;
end

>> y = myFunPlusOne(@sin,pi/2)
y = 2
>> y = myFunPlusOne(@cos,pi/2)
y = 1
>> y = myFunPlusOne(@sqrt,4)
y = 3
```

Using function handles requires that the function being assigned to a variable be saved in the current directory. An **anonymous function** is a function handle that is assigned to a function not stored. An anonymous function is created according to the following construction.

CONSTRUCTION: Anonymous function declaration.

```
F = @(inputs) function definition
```

TRY IT! Create myAdder from the example in the functions section using anonymous function handles. Use the variable F. Verify that F and myAdder behave the same for a = 1, b = 2, and c = 3.

```
>> F = @(a,b,c) a + b + c
>> y = F(1,2,3)
y = 6
>> y = myAdder(1,2,3)
y = 6
```

3.6 Script Files

A **script file** is an m-file that contains a sequence of instructions but is not a function. Unlike a function, a script file shares its workspace with the current directory. A script file is created by writing lines of code just as you would at the command prompt. A script file is **run** when you press the run button ▶ in the top toolbar. When the script file is run, MATLAB executes the instructions in order as if you typed them into the command prompt one at a time.

TRY IT! Write a script file that computes properties of a cylinder for a given radius and height.

```
clc
clear all
close all

% declare cylinder specs
r = 5;
h = 10;

% compute the volume of cylinder
V = pi*r^2*h

% compute lateral surface area of cylinder
s = 2*pi*r*h

% compute total surface area of cylinder
S = s + 2*pi*r^2
```

If you had a new cylinder, you could change r and h and rerun the script file to compute the new properties.

> **WARNING!** Script files share their workspace with the current directory, so be careful that variables assigned in the script file are not overwriting other variables with the same name.

> **TIP!** Always have a `clc`, `clear all`, and `close all` at the beginning of every script file. This is because the script shares the workspace with the current directory, and it is good practice to make sure that your script file is not using variables from old MATLAB sessions.

It is natural to wonder when a script file is more appropriate than a function or vice versa. Functions are most useful when a task needs to be done many times. Script files are useful when you need to perform a sequence of instructions only a few times in a highly context-specific situation. Some examples might be producing a complicated plot or trying something new to see if it is worth writing a function for it.

You can organize code in script files into **cells** (not to be confused with cell arrays), which can be run without running the rest of the script. This is not to be confused with cell arrays. A cell is a piece of script code that can be run individually. Code cells can be created by double commenting, or using two %%. You can execute a cell by clicking the evaluate-cell button or the evaluate-cell-and-advance button .

> **TRY IT!** Organize the previous script file into cells. Execute the cells one at a time using cell mode.

```
% Script File for Computing Properties of a Cylinder
clc
clear all
close all

% declare cylinder specs
r = 5;
h = 10;

% compute the volume of cylinder
V = pi*r^2*h

% compute lateral surface area of cylinder
s = 2*pi*r*h

% compute total surface area of cylinder
S = s + 2*pi*r^2
```

You can turn cell mode off or on under Cell → Disable Cell Mode.

Summary

1. A function is a self-contained set of instructions designed to do a specific task.
2. A function has its own workspace for its variables. Information can be added to a function's workspace only through a function's input variables. Information can leave the function's workspace only through a function's output variables.
3. You can assign functions to variables using function handles.
4. A script is a sequence of instructions that are executed in order. Variables in script files are shared with the working directory's workspace.
5. You can build your own functions and scripts in the MATLAB editor.

Vocabulary

anonymous function	function workspace	path
body of the function	input arguments	run
cell	keyword	script file
comment	local variable	subfunction
editor	m-file	type definition
function	output arguments	user
function handle	overload	
function header	parent function	

Functions and Operators

```
end                    return                 %%
function               %                      @
```

Problems

ṁ 1. Recall that the hyperbolic sine, denoted by sinh, is $\frac{\exp(x)-\exp(-x)}{2}$. Write a function with header [y] = mySinh(x), where y is the hyperbolic sine computed on x. Assume that x is a 1×1 double.

Test Cases:

```
>> y = mySinh(0)
y =

     0
>> y = mySinh(1)
y =
     1.1752
>> y = mySinh(2)
y =
     3.6269
```

ṁ 2. Write a function with header [M] = myCheckerBoard(n), where M is an $n \times n$ matrix with the following form:

$$M = \begin{bmatrix} 1 & 0 & 1 & 0 & 1 \\ 0 & 1 & 0 & 1 & 0 \\ 1 & 0 & 1 & 0 & 1 \\ 0 & 1 & 0 & 1 & 0 \\ 1 & 0 & 1 & 0 & 1 \end{bmatrix}$$

Note that the upper-left element should always be 1. Assume that n is a strictly positive integer.

Test Cases:

```
>> M = myCheckerboard(1)
M =

     1
>> M = myCheckerboard(2)
M =

     1     0
     0     1
>> M = myCheckerboard(3)
M =

     1     0     1
     0     1     0
```

```
         1       0       1
>> M = myCheckerboard(5)
M =
         1       0       1       0       1
         0       1       0       1       0
         1       0       1       0       1
         0       1       0       1       0
         1       0       1       0       1
```

.m 3. Write a function with header `[A] = myTriangle(b,h)` where A is the area of a triangle with base, b, and height, h. Recall that the area of a triangle is one-half the base times the height. Assume that b and h are 1×1 doubles.

Test Cases:
```
>> a = myTriangle(1,1)
a =
     0.5000
>> a = myTriangle(2,1)
a =
     1
>> a = myTriangle(12,5)
a =
    30
```

.m 4. Write a function with header `[M1,M2] = mySplitMatrix(M)`, where M is a matrix, M1 is the left half of M, and M2 is the right half of M. In the case where there is an odd number of columns, the middle column should go to M1. Hint: The `size` and `cell` functions will be useful for this. Assume that M has at least two columns.

Test Cases:
```
>> M = [1 2 3; 4 5 6; 7 8 9]
M =
         1       2       3
         4       5       6
         7       8       9
>> [M1, M2] = mySplitMatrix(M)
M1 =
         1       2
         4       5
         7       8
M2 =
     3
     6
     9
```

```
>> m = ones(5,5);
>> [m1,m2] = mySplitMatrix(m)
m1 =
     1     1     1
     1     1     1
     1     1     1
     1     1     1
     1     1     1
m2 =
     1     1
     1     1
     1     1
     1     1
     1     1
```

.m **5.** Write a function with header `[S,V] = myCylinder(r,h)`, where `r` and `h` are the radius and height of a cylinder, respectively, and `S` and `V` are the surface area and volume of the same cylinder, respectively. Recall that the surface area of a cylinder is $2\pi r^2 + 2\pi rh$, and the volume is $\pi r^2 h$. Assume that `r` and `h` are 1×1 doubles.

Test Cases:

```
>> [S,V] = myCylinder(1,5)
S =
   37.6991
V =
   15.7080
>> [S,V] = myCylinder(2,3)
S =
   62.8319
V =
   37.6991
```

.m **6.** Write a function with header `[N] = myNOdds(A)`, where `A` is a one-dimensional array of doubles and `N` is the number of odd numbers in `A`.

Test Cases:

```
>> n1 = myNOdds([1:100])
n1 =
    50
>> n2 = myNOdds([2:2:100])
n2 =
     0
```

.m **7.** Write a function with header `[out] = myTwos(m,n)`, where `out` is an $m \times n$ matrix of twos. Assume that `m` and `n` are strictly positive integers.

Test Cases:

```
>> A = myTwos(3,2)
A =
        2       2
        2       2
        2       2
>> A = myTwos(1,4)
A =
        2       2       2       2
```

[m] **8.** Write a function with header [S] = myAddString(S1,S2), where S is the concatenation of the strings S1 and S2.

Test Cases:

```
>> S = myAddString(myAddString('Programming', ' '), myAddString('is ', 'fun!'))
S = Programming is fun!
```

[m] **9.** Write a function that inputs a name, id, and grades, and generates a 1×1 struct array newStudent. Use the function to populate a 1×5 array called student.

● **10.** Generate the following errors:

```
function saved with incorrect name expects more input arguments.
output arguments not assigned
```

[m] **11.** Write a function with header [G] = myGreeting(name,age), where name is a string, age is a double, and G is the string 'Hi, my name is _____ and I am _____ years old.' where _____ are the input name and age, respectively. Hint: Use sprintf. Assume that name is a string and age is an integer.

Test Cases:

```
>> g1 = myGreeting('Timmy', 26)
g1 =
Hi, my name is Timmy and I am 26 years old.
>> g2 = myGreeting('Jill', 19)
g2 =
Hi, my name is Jill and I am 19 years old.
```

[m] **12.** Let r1 and r2 be the radius of circles with the same center and let r2>r1. Write a function with header [A] = myDonutArea(r1,r2), where A is the area outside of the circle with radius r1 and inside the circle with radius r2. Make sure that myDonutArea is vectorized. Assume that r1 and r2 are row vectors of the same size.

Test Case:

```
>> [ A ] = myDonutArea(1:5,2:2:10)
A =
     9.4248   37.6991   84.8230  150.7964  235.6194
```

m 13. Write a function with header `[indices] = myWithinTolerance(A, a, tol)`, where `indices` is an array of the indices in `A` such that $|A - a| <$ tol. Assume that `A` is a one-dimensional double array and that `a` and `tol` are 1×1 doubles.

Test Cases:
```
>> I = myWithinTolerance([0 1 2 3], 1.5, .75)
I =
      2      3

>> I = myWithinTolerance(0:.01:1, .5, .03)
I =
     48     49     50     51     52     53
```

m 14. Write a function with header `[boundedA] = myBoundingArray(A, top, bottom)`, where `boundedA` is equal to the array `A` wherever *bottom* $< A <$ *top*, `boundedA` is equal to `bottom` wherever $A <=$ *bottom*, and `boundedA` is equal to `top` wherever $A >=$ *top*. Assume that `A` is a one-dimensional double array and that `top` and `bottom` are 1×1 doubles.

Test Cases:
```
>> A = myBoundingArray(-5:5, 3, -3)
A =
    -3    -3    -3    -2    -1     0     1     2     3     3     3

>> x = linspace(0,2*pi,100);
>> plot(x, myBoundingArray(sin(x), .5, -.5))
>> axis([0 2*pi -1 1])
```

Branching Statements

CHAPTER OUTLINE

Motivation

When writing functions, it is very common to want certain parts of the function body to be executed only under certain conditions. For example, if the input argument is odd, you may want the function to perform one operation on it, and another if the input argument is even. This effect can be achieved in MATLAB using branching statements (i.e., the execution of the function *branches* under certain conditions), which are the topic of this chapter.

By the end of this chapter, you should be able to program branching statements into your functions and scripts, which should substantially increase the scope of tasks for which you will be able to make functions.

4.1 If-Else Statements

A **branching statement**, **If-Else Statement**, or **If-Statement** for short, is a code construct that executes blocks of code only if certain conditions are met. These conditions are represented as logical expressions. Let P, Q, and R be some logical expressions in MATLAB. The following shows an if-statement construction.

CONSTRUCTION: Simple If-Else Statement Syntax

```
if logical expression
    code block
end
```

The word "if" is a keyword. An if-statement is ended by the keyword end. When MATLAB sees an if-statement, it will determine if the associated logical expression is true. If it is true, then the code in code block will be executed by MATLAB. If it is false, then the code in the if-statement will not be executed. The way to read this is "If logical expression is true then do code block."

When there are several conditions to consider you can include elseif-statements; if you want a condition that covers any other case, then you can use an else statement.

CONSTRUCTION: Extended If-Else Statement Syntax

```
if logical expression P
    code block 1
elseif logical expression Q
    code block 2
elseif logical expression R
    code block 3
else
    code block 4
end
```

In the previous code, MATLAB will first check if P is true. If P is true, then code block 1 will be executed, and then *the if-statement will end*. In other words, MATLAB will *not* check the rest of the statements once it reaches a true statement. However, if P is false, then MATLAB will check if Q is true. If Q is true, then code block 2 will be executed, and the if-statement will end. If it is false, then R will be executed, and so forth. If P, Q, and R are all false, then code block 4 will be executed. You can have any number of elseif statements (or none) as long as there is at least one if-statement (the first statement). You do not need an else statement, but you can have at most one else statement. The logical expressions after the if and elseif (i.e., such as P, Q, and R) will be referred to as **conditional statements**.

TRY IT! Write a function with header [status] = myThermoStat(temp, desiredTemp). The value of status should be the string 'Heat' if temp is less than desiredTemp minus 5 degrees, 'AC' if temp is more than the desiredTemp plus 5, and 'off' otherwise.

```
function [status] = myThermoStat(temp,desiredTemp)
% [status] = myThermoStat(temp,desiredTemp)
% Changes status of thermostat based on temperature and desired temperature
% author
% date

if temp < desiredTemp - 5
    status = 'Heat';
elseif temp > desiredTemp + 5
    status = 'AC';
else
    status = 'off';
end
```

```
end % end myThermoStat


>> status = myThermoStat(65, 75)
status =
     Heat
>> status = myThermoStat(75, 65)
     AC
>> status = myThermoStat(65, 63)
status =
     off
```

EXAMPLE: What will be the value of y after the following script is executed?

```
x = 3;
if x > 1
    y = 2;
elseif x > 2
    y = 4;
else
    y = 0;
end
    >> y
    y =
        2
```

We can also insert more complicated conditional statements using logical operators.

EXAMPLE: What will be the value of y after the following code is executed?

```
x = 3;
if x > 1 && x < 2
    y = 2;
elseif x > 2 && x < 4
    y = 4;
else
    y = 0;
end


>> y
y =
    4
```

> **WARNING!** Remember, if you want the logical statement $a < x < b$, this is *two* conditional statements, $a < x$ AND $x < b$. Typing a < x < b will have unexpected and undesirable results.

A statement is called **nested** if it is entirely contained within another statement of the same type as itself. For example, a **nested if-statement** is an if-statement that is entirely contained within a clause of another if-statement.

> **EXAMPLE:** Think about what will happen when the following code is executed. What are all the possible outcomes based on the input values of x and y?
>
> ```
> function [out] = myNestedBranching(x,y)
> % [out] = myNestedBranching(x,y)
> % Nested Branching Statement Example
> % author
> % date
>
> if x > 2
>
>     if y < 2
>         out = x + y;
>     else
>         out = x - y;
>     end
>
> else
>
>     if y > 2
>         out = x*y;
>     else
>         out = 0;
>     end
>     end
>     end
> ```

> **TIP!** To help keep track of which code blocks belong under which conditional statement, MATLAB gives the same level of indentation to every line of code within a conditional statement. As you write code, you may find that the indentation becomes incorrect for whatever reason. You can indent everything properly by pressing ctrl+a to select all your code and then ctrl+i to properly indent on a PC, and command+a and command+i on a MAC. Be sure to use this sequence of instructions on your code before presenting it to other people. It makes it much easier to read.

EXAMPLE: Following is the code for myNestedBranching without any indentation. Not only is it not visually pleasing, but it also makes it much harder to understand the structure of the code.

```
function [out] = myNestedBranching(x,y)
% [out] = my_NestedBranching(x,y)
% Nested Branching Statement Example
% author
% date

if x > 2
if y < 2
out = x + y;
else
out = x - y;
end
else
if y > 2
out = x*y;
else
out = 0;
end
end
end
```

TIP! When learning to program to it is natural write code from beginning to end, just the way you write sentences. However, it is usually better to write complete if-statements first (all the conditional statements) before beginning to fill in the code block sections. For example, when writing an if-statement, write the "if" at the top, then the "end" at the bottom, then fill in the elseif and else statements, then fill in the body of each individual statement. Although it is trivial for the examples given in this chapter, coding in this order will help you keep track of your code when it becomes more complicated (as it will later in the book).

EXAMPLE: The following shows a good order in which to type myNestedBranching.

Step 1: Declare the function header and comments.

```
function [out] = myNestedBranching(x,y)
% [out] = myNestedBranching(x,y)
% Nested Branching Statement Example
% author
% date


end % end myNestedBranching
```

Step 2: Write main branching statement (outermost if-statement first).

```
function [out] = myNestedBranching(x,y)
% [out] = myNestedBranching(x,y)
% Nested Branching Statement Example
% author
% date

if x > 2


else


end


end % end myNestedBranching
```

Step 3: Fill in the code block for the first conditional statement (i.e., the nested if-statement).

```
function [out] = myNestedBranching(x,y)
% [out] = myNestedBranching(x,y)
% Nested Branching Statement Example
% author
% date

if x > 2

    if y < 2
        out = x + y;
    else
        out = x - y;
    end

else



end


end % end myNestedBranching
```

Step 4: Fill in the code block for the second conditional statement.

```
function [out] = myNestedBranching(x,y)
% [out] = myNestedBranching(x,y)
% Nested Branching Statement Example
% author
% date

if x > 2

    if y < 2
        out = x + y;
    else
        out = x - y;

        end
else

    if y > 2
        out = x*y;
    else
        out = 0;
    end

end

end % end myNestedBranching
```

Writing code in this way will help you break down your task in a way that will help you program it effectively.

There are many logical functions that are designed to help you build branching statements. For example, you can ask if a variable has a certain data type or value with functions like isreal, isnan, isinf, and isa. There are also functions that can tell you information about arrays of logicals like any, which computes to true if any element in an array is true, and false otherwise, and all, which computes to true only if all the elements in an array are true.

Sometimes you may want to design your function to check the inputs of a function to ensure that your function will be used properly. For example, the function myAdder in the previous chapter expects doubles as input. If the user inputs a struct or a char as one of the input variables, then the function will throw an error or have unexpected results. To prevent this, you can put a check to tell the user the function has not been used properly. This and other techniques for controlling errors are explored further in Chapter 9. For the moment, you only need to know that the error function stops a function's execution and throws an error with the text in the input string. The error function takes sprintf type inputs.

EXAMPLE: Modify `myAdder` to throw an error if the user does not input doubles. Try your function for nondouble inputs to show that the check works.

```
function [out] = myAdder(a,b,c)
% [out] = myAdder(a,b,c)
% out is the sum of a,b, and c.
% author
% date

% check for erroneous input
if ~isa(a, 'double') || ~isa(b, 'double') || ~isa(c, 'double')
    error('Inputs a, b, and c must be doubles.')
end

% assign output
out = a + b + c;

end


>> x = myAdder(1,2,3)
x = 6;

>> y = myAdder('1','2','3')
Error: myAdder inputs a, b, and c must be doubles.
```

There is a large variety of erroneous inputs that your function may encounter from users, and it is unreasonable to expect that your function will catch them all. Therefore, unless otherwise stated, write your functions assuming the functions will be used properly.

The remainder of the section gives a few more examples of branching statements.

TRY IT! Write a function called `isOdd` that returns `'odd'` if the input is odd and `'even'` if it is even. You can assume that `in` will be a positive integer.

```
function [out] = isOdd(in)
% [out] = isOdd(in)
% returns out = 'odd' if in is odd, 'even' if in is even
% author
% date

% use remainder function to check if in is divisible by 2.
if rem(in,2) == 0
    % if it is divisible by 2, then in is not odd
```

```
     out = 'even';
else
     out = 'odd';
end % end if rem

end % end isOdd
```

TRY IT! Write a function called `myCircCalc` that takes a double, `r`, and a string, `calc` as input arguments. You may assume that `r` is positive, and that `calc` is either the string `'area'` or `'circumference'`. The function `myCircCalc` should compute the area of a circle with radius, `r`, if the string `calc` is `'area'`, and the circumference of a circle with radius, `r`, if `calc` is `'circumference'`. It will be helpful to use the MATLAB function `strcmp`. The function should be vectorized for `r`.

```
function [out] = myCircCalc(r, calc)
% [out] = myCircCalc(r, calc)
% returns out = pi*r^2 if calc is 'area' and out = 2*pi*r if calc is 'circumference'
% author
% date
if strcmp(calc, 'area')
    out = pi*r.^2;
elseif strcmp(calc, 'circumference')
    out = 2*pi*r;
end

end % end myCircCalc
```

Summary

1. Branching (if-else) statements allow functions to take different actions under different circumstances.

Vocabulary

branching statement	if-statement	nested if-statement
conditional statement	nested	
if-else statement		

Functions and Operators

all	Elsie	isnan
any	if	isreal
ctrl+i	isa	
else	isinf	

Problems

.m **1.** Write a function with header [tip] = myTipCalc(bill, party), where bill is the total cost of a meal and party is the number of people in the group. The tip should be calculated as 15% for a party strictly less than six people, 18% for a party strictly less than eight, 20% for a party less than 11, and 25% for a party 11 or more.

Test Cases:

```
>> t = myTipCalc(109.29,3)
t =
    16.3935
>> t = myTipCalc(109.29,7)
t =
    19.6722
>> t = myTipCalc(109.29,9)
t =
    21.8580
>> t = myTipCalc(109.29,12)
t =
    27.3225
```

.m **2.** Write a function with header [f] = myMultOperation(a,b,operation). The input argument, operation, is a string that is either 'plus', 'minus', 'mult', 'div', or 'pow', and f should be computed as a+b, a-b, a*b, a/b, and a^b for the respective values for operation. Be sure to make your function vectorized. Hint: Use the strcmp function.

Test Cases:

```
>> x = [1 2 3 4];
>> y = [2 3 4 5];
>> f = myMultOperation(x,y,'plus')
f =
      3     5     7     9
>> f = myMultOperation(x,y,'minus')
f =
     -1    -1    -1    -1
>> f = myMultOperation(x,y,'mult')
f =
      2     6    12    20
>> f = myMultOperation(x,y,'div')
```

```
f =
     0.5000     0.6667     0.7500     0.8000
>> f = myMultOperation(x,y,'pow')
f =
          1          8         81       1024
```

.m 3. Consider a triangle with vertices at (0,0), (1,0), and (0,1). Write a function with header `[S] = myInsideTriangle(x,y)` where S is the string `'outside'` if the point `(x,y)` is outside of the triangle, `'border'` if the point is exactly on the border of the triangle, and `'inside'` if the point is on the inside of the triangle.

Test Cases:

```
EDU>> S = myInsideTriangle(.5,.5)
S =
border
EDU>> S = myInsideTriangle(.25,.25)
S =
inside
EDU>> S = myInsideTriangle(5,5)
S =
outside
```

.m 4. Write a function with header `[out] = myMakeSize10(x)`, where x is an array and out is the first 10 elements of x if x has more than 10 elements, and out is the array x padded with enough zeros to make it length 10 if x has less than 10 elements.

Test Cases:

```
>> A = myMakeSize10(1:2)
A =
     1     2     0     0     0     0     0     0     0     0
>> A = myMakeSize10(1:15)
A =
     1     2     3     4     5     6     7     8     9    10
 >> A = myMakeSize10([3 6 13 4])
A =
     3     6    13     4     0     0     0     0     0     0
```

5. Can you write `myMakeSize10` without using if-statements (i.e., using only logical and array operations)?

.m 6. Write a function with header `[grade] = myLetterGrader(percent)`, where grade is the string `'A+'` if percent is greater than 97, `'A'` if percent is greater than 93, `'A-'` if percent is greater than 90, 'B+' if percent is greater than 87, 'B' if percent is greater than 83, 'B-' if percent is greater than 80, 'C+' if percent is greater than 77, 'C' if percent is greater than 73, 'C-' if percent is greater than 70, 'D+' if percent is greater than 67, 'D' if percent is greater than 63, 'D-' if percent is greater than 60, and 'F' for any percent less than 60. Grades exactly on the division should be included in the higher grade category.

Test Cases:

```
>> grade = myLetterGrader(97)
grade =
A+
>> grade = myLetterGrader(84)
grade =
B
```

⌞m 7. Most engineering systems have redundancy. That is, an engineering system has more than is required to accomplish its purpose. Consider a nuclear reactor whose temperature is monitored by three sensors. An alarm should go off if any two of the sensor readings disagree. Write a function with header [response] = myNukeAlarm(S1,S2,S3) where S1, S2, and S3 are the temperature readings for sensor 1, sensor 2, and sensor 3, respectively. The output response should be the string 'alarm!' if any two of the temperature readings disagree by strictly more than 10 degrees and 'normal' otherwise.

Test Cases:

```
>> response = myNukeAlarm(94,96,90)
response =
normal
>> response = myNukeAlarm(94,96,80)
response =
alarm!
>> response = myNukeAlarm(100,96,90)
response =
normal
```

⌞m 8. Let $Q(x)$ be the quadratic equation $Q(x) = ax^2 + bx + c$ for some scalar values a, b, and c. A root of $Q(x)$ is an r such that $Q(r) = 0$. The two roots of a quadratic equation can be described by the quadratic formula, which is

$$r = \frac{-b \pm \sqrt{b^2 - 4ac}}{2a}.$$

A quadratic equation has either two real roots (i.e., $b^2 > 4ac$), two imaginary roots (i.e., $b^2 < 4ac$), or one root, $r = -\frac{b}{2a}$.

Write a function with header [nRoots, r] = myNRoots(a,b,c) where a, b, and c are the coefficients of the quadratic $Q(x)$, nRoots is 2 if Q has two real roots, 1 if Q has one root, -2 if Q has two imaginary roots, and r is an array containing the roots of Q.

Test Cases:

```
>> [nRoots, r] = myNRoots(1,0,-9)
nRoots =
        2
r =
        3     -3
>> [nRoots, r] = myNRoots(3,4,5)
nRoots =
       -2
r =
   -0.6667 + 1.1055i   -0.6667 - 1.1055i

>> [nRoots, r] = myNRoots(2,4,2)
nRoots =
        1
r =
       -1
```

.m 9. Write a function with header [h] = mySplitFunction(f,g,a,b,x), where f and g are handles to functions f(x) and g(x), respectively. The output argument h should be f(x) if x ≤ a, g(x) if x ≥ b, and 0 otherwise. You may assume that b > a.

Test Cases:

```
>> h = mySplitFunction(@exp, @sin, 2, 4, 1)
h =
      2.7183
>> h = mySplitFunction(@exp, @sin, 2, 4, 3)
h =
      0
>> h = mySplitFunction(@exp, @sin, 2, 4, 5)
h =
     -0.9589
```

Iteration

5

Motivation

Many tasks in life are boring or tedious because they require doing the same basic actions over and over again—iterating—in slightly different contexts. For example, consider looking up the definition of 20 words in the dictionary, populating a large table of numbers with data, alphabetizing many stacks of paper, or dusting off every object and shelf in your room. Since repetitive tasks appear so frequently, it is only natural that programming languages like MATLAB would have direct methods of performing iteration.

This chapter teaches you how to program iterative tasks. With branching and iteration, it is possible to program just about any task that you can imagine.

5.1 For-Loops

A **for-loop** is a sequence of instructions that is repeated, or iterated, for every value of a **looping array**. The variable that holds the current value of the looping array is called **looping variable**. Sometimes for-loops are referred to as **definite loops** because they have a predefined begin and end. The abstract syntax of a for-loop block is as follows.

CONSTRUCTION: For-loop

```
for looping variable = looping array
    code block
end
```

A for-loop assigns the looping variable to the first element of the looping array. It executes everything in the code block. Then it assigns the looping variable to the next element of the looping array and

An Introduction to MATLAB® Programming and Numerical Methods. http://dx.doi.org/10.1016/B978-0-12-420228-3.00005-1

executes the code block again. It continues this process until there are no more elements in the looping array to assign.

TRY IT! What is the sum of every integer from 1 to 3?

```
N = 0;
for i = 1:3
    N = N + i;
end
```

WHAT IS HAPPENING? First, recall that `1:3` is the array `[1 2 3]`.

1. The variable `N` is assigned the value 0.
2. The variable `i` is assigned the value 1 (the first element in the array `[1:3]`).
3. The variable `N` is assigned the value `N + i` ($0 + 1 = 1$).
4. The variable `i` is assigned the value 2.
5. The variable `N` is assigned the value `N + i` ($1 + 2 = 3$).
6. The variable `i` is assigned the value 3.
7. The variable `N` is assigned the value `N + i` ($3 + 3 = 6$).
8. With no more values to assign in the looping array, the for-loop is terminated with `N = 6`.

We present several more examples to give you a sense of how for-loops work, even though some of these examples could be performed more concisely with array operations.

EXAMPLE: Given a one-dimensional array of strictly positive integers, `A`, add all the elements of `A`. Recall that the length of an array can be determined by using the `length` function. You may assume that `A` is defined in the current workspace.

```
S = 0;
for i = 1:length(A)
    S = S + A(i);
end
```

Abstractly, `i` represents the index of `A` that we are adding to our total sum. `S` represents the sum of the elements of `A` up to index `i`. For `A = [4 1 2]`, go through the code step by step and note every action executed by the computer.

The MATLAB function sum has already been written to handle the previous example. However, assume you wish to add only the even numbers of an array. What would you add to the previous for-loop block to handle this restriction?

EXAMPLE: Given a one-dimensional array of strictly positive integers, A, add only the even elements of A. You may assume that A is defined in the current workspace. Recall that a number is even if it can be divided by 2 without remainder.

```
S = 0;
for i = 1:length(A)
    if rem(A(i),2) == 0
        S = S + A(i);
    end
end
```

The remainder function, rem(a,b), returns the remainder of a number divided by b. The logical statement if rem(A(i),2) == 0 asks "is the remainder of A(i) divided by 2 equal to 0?" If the result is true, then A(i) is added to the total sum. If the result is false, then the code within the if-statement is skipped. For A = [1 2 3 4 5], go through the code step by step and note every action executed by the computer as in the previous example. What would you change in the code so that only odd numbers in A are added? Hint: You can do it by adding only a single character.

EXAMPLE: Let the function anyEs have type S (string) → out (logical). The output variable, out, should take the value 1 if there are any E's in the input string, S, and 0 otherwise. Assume every character in S is lowercase.

```
function [out] = anyEs(S)
% [out] = anyEs(S)
% out is 1 if and only if there is an 'E' in the input string, S. S is assumed to be lowercase.

% start by assuming there are no E's in S
out = 0;

% check every letter in S, S(i), for the value 'E'
% i is the ith letter in S
for i = 1:length(S)
    if strcmp(S(i),'e')

        % if letter i is an 'E', then there is at least one 'E' in S
        out = 1;

        % since we are only checking for any E's, we can stop looking if we find one
        break
    end
end % end for i

end % end anyEs
```

The first step in the function anyEs assumes that there are no E's in S (i.e., the output is 0 or false). This is not a trivial assumption, because we could just as easily have started out assuming that there was an E in S (i.e., output is true). The reason it is easier to start out assuming that there are no E's in S is because if we ever find one, then we know our original assumption was wrong and the output should be true and we can stop looking. However, if we start out assuming there is an E in S, then if we find a character that is not an E, it says nothing about whether or not our original assumption was true, and we have to keep looking. As an exercise, try to rewrite `anyEs` starting with the assumption that there is an E in S (`out = 1`). Was it easier or harder? Programming experience will help you determine which assumptions will make your programming task easier. The program also uses the function `strcmp`, which computes to 1 if the two input strings are the same and 0 otherwise.

Notice the new keyword `break`. If executed, the `break` keyword immediately stops the most immediate for-loop that contains it; that is, if it is contained in a nested for-loop, then it will only stop the innermost for-loop. In this particular case, the `break` command is executed if we ever find an E. The code will still function properly without this statement, but since the task is to find out if there are *any* E's in S, we do not have to keep looking if we find one. Similarly, if a human was given the same task for a long string of characters, that person would not continue looking for E's if he or she already found one. Break statements are used when anything happens in a for-loop that would make you want it to stop early. A less intrusive command is the keyword `continue`, which skips the remaining code in the current iteration of the for-loop, and continues on to the next element of the looping array.

TIP! It is generally good practice to label the end of a for-loop if it is many lines away from its beginning. In the function `anyEs`, the end of the for-loops is labeled with the comment `% end for i`. Additionally, it is often helpful to place a comment before the for-loop stating what the for-loop index represents and what the for-loop does.

EXAMPLE: Let the function `myDist2Points` have header `[d] = myDist2Points (XY,xy)`. The input argument `XY` is a two-column array where each row denotes the x-y coordinates of a point in Euclidean space, `xy` is a 1×2 array containing an x-y coordinate, and `d` is an array containing the distances from `xy` to the points contained in each row of `XY`. Write the function `myDist2Points` in MATLAB.

```
function [d] = myDist2Points(XY,xy)
% [d] = myDist2Points(XY,xy)
% Returns a list of distances between xy and the points contained in the rows of XY
% author
% date

% initialize empty array of distances
d = zeros(1,nRows);

% loop through the rows of XY
% i is the current row
for i = 1:size(XY,1)
    % add this distance to the list of distances
    d = [d, sum((XY(i,:) - xy).^2)];
end % end for i
end % end myDist2Points
```

WARNING! If you create this function and save it as an m-file, MATLAB will give you a warning for the line d = [d, sqrt(sum((XY(i,:) - xy).^2))]; by underlining it yellow. This warning is given because the variable d is getting larger at every iteration of the for-loop, which, due to reasons beyond the scope of this book, is very time consuming for the computer. For most cases, the reduction in speed is not noticeable, but it can be a problem if it is done many times for large data sets (on the order of tens of thousands). If you know what the length the array will be when the for-loop finishes, you can save time by **preallocating** the elements of the array. That is, you can set d to an array of the proper size with dummy values (usually 0) for every element. The previous code is rewritten with d preallocated.

```
function [d] = myDist2Points(XY,xy)
% [d] = myDist2Points(XY,xy)
% Returns a list of distances between xy and the points contained in the rows of XY
% author
% date

% get number of rows in XY
nRows = size(XY,1);

% preallocate empty array of distances
d = zeros(1; nRows);

% loop through the rows of XY, is the current row
for i = 1:
    % add this distance to the list of distances
    d(i) = sum((XY(i,:) - xy).^2);
end % end for i
end % end myDist2Points
```

Just like if-statements, for-loops can be nested.

EXAMPLE: Let A be a two-dimensional matrix, [5 6; 7 8], defined in the current workspace. Use a nested for-loop to sum all the elements in A. How would you adapt this code to handle A of arbitrary size?

```
S = 0;
for i = 1:2
    for j = 1:2
        S = S + A(i,j);
    end

    end

    S =
        26
```

WHAT IS HAPPENING?

1. S, representing the running total sum, is set to 0.
2. The outer for-loop begins with looping variable, `i`, set to 1.
3. Inner for-loop begins with looping variable, `j`, set to 1.
4. S is incremented by `A(i,j) = A(1,1) = 5`. So S = 5.
5. Inner for-loop sets `j = 2`.
6. S is incremented by `A(i,j) = A(1,2) = 6`. So S = 11.
7. Inner for-loop terminates.
8. Outer for-loop sets `i = 2`.
9. Inner for-loop begins with looping variable, `j`, set to 1.
10. S is incremented by `A(i,j) = A(2,1) = 7`. So S = 18.
11. Inner for-loop sets `j = 2`.
12. S is incremented by `A(i,j) = A(2,2) = 8`. So S = 26.
13. Inner for-loop terminates.
14. Outer for-loop terminates with S = 26.

WARNING! Although possible, do not try to change the looping variable inside of the for-loop. It will make your code very complicated and will likely result in errors.

EXAMPLE: Find the sum of the squares of every number from 1 to 10. Two examples are given. The first is *improper*. The second example is recommended.

Improper implementation:

```
S = 0;
for i = 1:10
    i = i^2;
    S = S + i;
end
```

Proper implementation:

```
S = 0;
for i = 1:10
    S = S + i^2;
end
```

TRY IT! In Chapter 3 we created the function `myAdder`, which adds three input scalars. Now assume we are given a list of number triplets in the form of a three-column array, A (i.e., each row in A is a triplet). The number of rows in the array will not be specified. Write a function with header

[out] = myListAdder(A), where out is a single column where each element is the output of myAdder for the corresponding row in A. Your function should call myAdder and use a single for-loop.

```
function [out] = myListAdder(A)
% [out] = myListAdder(A)
% Returns a column vector out where out(i) = myAdder(A(i,1), A(i,2), A(i,3))
% author
% date

% get size of A
[m,n] = size(A);

% preallocate out
out = zeros(m,1);

% i is the row in A
for i = 1:m
    out(i) = myAdder(A(i,1), A(i,2), A(i,3));
end % end for i

end % end myListAdder
```

5.2 Indefinite Loops

A **while-loop** or an **indefinite loop** is a set of instructions that is repeated as long as the associated logical expression is true. The abstract syntax of a while loop block is the following:

CONSTRUCTION: While-loop

```
while logical expression
    code block
end
```

When MATLAB reaches a while-loop block, it first determines if the logical expression of the while-loop is true or false. If the expression is true, then the code block will be executed. If the expression is false, then the while-loop will terminate.

TRY IT! Determine the number of times 8 can be divided by 2 until the result is less than 1.

```
n = 0;
N = 8;
while N >= 1
    N = N/2;
    n = n + 1;
end
```

WHAT IS HAPPENING?

1. First the variable n (running count of divisions of N by 2) is set to 0.
2. N is set to 8 and represents the current value we are dividing by 2.
3. The while-loop begins.
4. MATLAB computes N >= 1 or 8 >= 1, which is true so the code block is executed.
5. N is assigned N/2 = 8/2 = 4.
6. n is incremented to 1.
7. MATLAB computes N >= 1 or 4 >= 1, which is true so the code block is executed.
8. N is assigned N/2 = 4/2 = 2.
9. n is incremented to 2.
10. MATLAB computes N >= 1 or 2 >= 1, which is true so the code block is executed.
11. N is assigned N/2 = 2/2 = 1.
12. n is incremented to 3.
13. MATLAB computes N >= 1 or 1 >= 1, which is true so the code block is executed.
14. N is assigned N/2 = 1/2 = 0.5.
15. n is incremented to 4.
16. MATLAB computes N >= 1 or 0.5 >= 1, which is false so the while-loop ends with n = 4.

TRY IT! A fun way to test the speed of your computer is to see how high it can count in one second. Note that the function `tic` starts an internal timer in MATLAB. The function `toc` returns the time in seconds since the last `tic`. As of the writing of this text, a typical PC laptop can count to about 100,000, and a MAC can count to about a million. How high can a human get? Hopefully this little test can give you a sense of how fast computers are compared to humans.

```
N = 0;
tic
while toc <= 1
    N = N + 1;
end
N
```

WHAT IS HAPPENING?

1. The variable N (current count) is set to 0.
2. The variable A timer is started using the tic function.
3. The `while` loop begins.
4. MATLAB computes the time since the last `tic` using the `toc` function.
5. MATLAB checks if the time is less than 1 second, which is probably true on the first iteration.
6. The variable N is incremented by 1.
7. MATLAB computes the time since the last `tic` using the `toc` function.

8. MATLAB checks if the time is less than 1 second, which is probably true on the second iteration.
9. The variable N is incremented by 1.
10. ...
11. This process is repeated until `toc` returns a time greater than 1 second.
12. The `while` loop ends with N as the final count.

You may have asked, "What if the logical expression is true and never changes?" and this is indeed a very good question. If the logical expression is true, and nothing in the while-loop code changes the expression, then the result is known as an **infinite loop**. Infinite loops run forever, or until your computer breaks or runs out of memory.

EXAMPLE: Write a while-loop that causes an infinite loop.

```
N = 0;
while N > −1
    N = N + 1;
end
```

Since N will always be bigger than −1 no matter how many times the loop is run, this code will never end. Can you change a single character so that the while-loop will run at least once but will not infinite loop?

Infinite loops are not always easy to spot. Consider the next two examples: one infinite loops and one does not. Can you determine which is which? As your code becomes more complicated, it will become harder to detect.

EXAMPLE: Which while-loop causes an infinite loop? Recall that the `rem(a,b)` function returns the remainder of a/b.

```
N = 1;
while N > 0
    N = N/2;
end


N = 2;
while N > 0
    if rem(N,2) == 0
        N = N + 1;
    else
        N = N − 1;
    end
end
```

Answer: The first example will not infinite loop because eventually N will be so small that MATLAB cannot tell the difference between N and 0. More on this in Chapter 8. The second example will infinite loop because N will oscillate between 2 and 3 indefinitely.

What happens in the case of an infinite loop? Do you have to force MATLAB to shut down, or restart your computer, or buy a new one? Fortunately MATLAB allows you to stop any code by pressing ctrl+c. If you think your code is stuck in an infinite loop, or if you are just tired of waiting for it to do its job, you can use this command to force your code to stop. An error will be thrown at the line that was being executed when ctrl+c was executed.

Summary

1. Loops provide a mechanism for code to perform repetitive tasks; that is, iteration.
2. There are two kinds of loops: for-loops and while-loops.
3. Loops are important for constructing iterative solutions to problems.

Vocabulary

break	indefinite loop	preallocate
continue	infinite loop	while-loop
definite loop	looping array	
for-loop	looping variable	

Functions and Operators

break	ctrl+c	while
continue	for	

Problems

1. What will the value of y be after the following code is executed?

```
y = 0;
for i = 1:1000
    for j = 1:1000
        if i==j
            y = y + 1;
        end
    end
end
```

⌐m 2. Write a function with header [M] = myMax(A) where M is the maximum (largest) value in A. Do not use the built-in MATLAB function max.

⌐m 3. Write a function with header [M] = myNMax(A,N) where M is an array consisting of the N largest elements of A. You may use MATLAB's max function. You may also assume that N is less than the length of M, that A is a one-dimensional array with no duplicate entries, and that N is a strictly positive integer smaller than the length of A.

Test Case:

```
>> x = [7,9,10,5,8,3,4,6,2,1];
>> M = myNMax(x,  3)
M =
       10      9      8
```

⌐m 4. Let M be a matrix of positive integers. Write a function with header [Q] = myTrigOddEven(M), where $Q(i, j) = \sin(M(i, j))$ if $M(i, j)$ is even, and $Q(i, j) = \cos(M(i, j))$ if $M(i, j)$ is odd.

⌐m 5. Let P be an $m \times p$ matrix and Q be a $p \times n$ matrix. As you will find later in this book, $M = P \times Q$ is defined as $M(i, j) = \sum_{k=1}^{p} P(i, k) \cdot Q(k, j)$. Write a function with header [M] = myMatMult(P,Q) that uses for-loops to compute M, the matrix product of P and Q. Hint: You may need up to three nested for-loops.

Test Cases:

```
>> P = ones(3,3);
>> M = myMatMult(P,P)
M =
       3      3      3
       3      3      3
       3      3      3

>> P = [1 2 3 4; 5 6 7 8];
>> Q = [1 1 1; 2 2 2; 3 3 3; 4 4 4];
>> M = myMatMult(P,Q)
M =
      30     30     30
      70     70     70
```

⌐m 6. The interest, i, on a principle, P_0, is a payment for allowing the bank to use your money. Compound interest is accumulated according to the formula $P_n = (1 + i)P_{n-1}$, where n is the compounding period, usually in months or years. Write a function with header [years] = mySavingPlan(P0, i, goal) where years is the number of years it will take P_0 to become goal at $i\%$ interest compounded annually.

Test Cases:

```
>> y = mySavingPlan(1000, 0.05, 2000)
y =
     15
>> y = mySavingPlan(1000, 0.07, 2000)
y =
     11
>> y = mySavingPlan(500, 0.07, 2000)
y =
     21
```

.m **7.** Write a function with header [ind] = myFind(B), where ind is an array of indices i where B(i) is 1. You may assume that B is an array of only ones and zeros. Do not use the built-in MATLAB function find.

.m **8.** Assume you are rolling two six-sided dice, each side having an equal chance of occurring. Write a function with header [roll] = myMonopolyDice(), where roll is the sum of the values of the two dice thrown but with the following extra rule: if the two dice rolls are the same, then another roll is made, and the new sum added to the running total. If the two dice show 3 and 4, then the running total should be 7. If the two dice show 1 and 1, then the running total should be 2 plus the total of another throw. Rolls stop when the dice rolls are different. Hint: The line result = randi([1 6],2,1) is an accurate simulation of rolling two dice (See Figure 5.1).

Test Cases:

```
>> rolls = zeros(1,100000);
>> for i = 1:100000
rolls(i) = myMonopolyDice();
    end
>> hist(rolls, 25)
>> xlabel('Roll Value')
>> ylabel('Number of Occurences')
>> title('Histogram of Monopoly Dice Rolls')
```

.m **9.** A number is prime if it is divisible without remainder only by itself and 1. The number 1 is not prime. Write a function with header [out] = myIsPrime(n), where out is 1 if n is prime and 0 otherwise. Assume that n is a strictly positive integer.

.m **10.** Write a function with header [primes] = myNPrimes(N) where primes is a list of the first N primes. Assume that N is a strictly positive integer.

.m **11.** Write a function with header [fibPrimes] = myNFibPrimes(N), where fibPrimes is a list of the first N numbers that are both a Fibonacci number and prime. Note: 1 is not prime. Hint: Do not use the recursive implementation of Fibonacci numbers. A function to compute Fibonacci numbers is presented in Section 6.1. You may use the code freely.

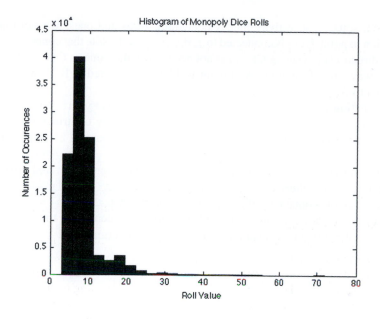

FIGURE 5.1

Note: Histogram may look slightly different based on your computer.

Test Cases:

```
>> FP = myNFibPrimes(3)
FP =
      3     5    13
>> FP = myNFibPrimes(8)
FP =

      3     5      13      89     233    1597   28657   514229
```

.m 12. Write a function with header `[Q] = myTrigOddEven(M)`, where $Q(i, j) = \sin(\pi/M(i, j))$ if `M(i,j)` is odd, and $Q(i, j) = \cos(\pi/M(i, j))$ if `M(i,j)` is even. Assume that `M` is a two-dimensional matrix of strictly positive integers.

Test Cases:

```
>> M = [3 4; 6 7];
>> Q = myTrigOddEven(M)
Q =
    0.8660    0.7071
    0.8660    0.4339
```

.m 13. Let C be a square connectivity matrix containing zeros and ones. We say that point i has a connection to point j, or i is connected to j, if $C(i, j) = 1$. Note that connections in this context are one-directional, meaning $C(i, j)$ is not necessarily the same as $C(j, i)$. For example, think of a one-way street from point A to point B. If A is connected to B, then B is not necessarily connected to A.

Write a function with header [node] = myConnectivityMat2Struct(C, names), where C is a connectivity matrix and names is a cell array of strings (i.e., each element of names is a string) that denote the name of a point. That is, names(i) is the name of the $i - th$ point.

The output variable node should be a struct with fields .name and .neighbors. The $i - th$ element of node is defined as node(i).name = names(i) and node(i).neighbors is a row vector containing the indices, j, such that C(i,j) = 1. In other words, node(i).neighbors is a list of points that point i is connected to.

Warning: Make sure the field names are exactly correct: .name and .neighbors.

Test Cases:

```
>> C = [0 1 0 1; 1 0 0 1; 0 0 0 1; 1 1 1 0]
C =
        0     1     0     1
        1     0     0     1
        0     0     0     1
        1     1     1     0
>> names =  {'Los Angeles', 'New York', 'Miami', 'Dallas'};
>> node = myConnectivityMat2Struct(C, names);
>> node(1)
ans =
         name: 'Los Angeles'
    neighbors: [2 4]
>> node(2)
ans =
         name: 'New York'
    neighbors: [1 4]
>> node(3)
ans =
         name: 'Miami'
    neighbors: 4
>> node(4)
ans =
         name: 'Dallas'
    neighbors: [1 2 3]
```

Recursion

6

CHAPTER OUTLINE

Motivation

Imagine that a CEO of a large company wants to know how many people work for him. One option is to spend a tremendous amount of personal effort counting up the number of people on the payroll. However, the CEO has other more important things to do, and so implements another, more clever, option. At the next meeting with his department directors, he asks everyone to tell him at the next meeting how many people work for them. Each director then meets with all their managers, who subsequently meet with their supervisors who perform the same task. The supervisors know how many people work under them and readily report this information back to their managers (plus one to count themselves), who relay the aggregated information to the department directors, who relay the relevant information to the CEO. In this way, the CEO accomplishes a difficult task (for himself) by delegating similar, but simpler, tasks to his subordinates.

This method of solving difficult problems by breaking them up into simpler problems is naturally modeled by recursive relationships, which are the topic of this chapter, and which form the basis of important engineering problem-solving techniques. By the end of this chapter, you should be able to recognize recursive relationships and program them using recursive functions.

6.1 Recursive Functions

A **recursive** function is a function that makes calls to itself. Although a recursive function is defined in terms of itself, MATLAB opens a new workspace every time a function is called, even for a function calling a function with the same name as itself.

In recursive functions, the **base case** is the mechanism that stops it from calling itself indefinitely. The base case is usually an input value for which there is an easily verifiable solution. The **recursive step** is the set of all cases where a **recursive call**, or a function call to itself, is made.

An Introduction to MATLAB® Programming and Numerical Methods. http://dx.doi.org/10.1016/B978-0-12-420228-3.00006-3

As an example, we show how recursion can be used to define and compute the factorial of an integer number. The factorial of an integer number, n, is $1 \times 2 \times 3 \times \cdots \times (n-1) \times n$. The factorial function can also be defined as n times the factorial of $n - 1$. This recursive definition can be written:

$$f(n) = \begin{cases} 1 & \text{if } n = 1 \\ n \times f(n-1) & \text{otherwise} \end{cases}$$

The base case is $n = 1$ for which the factorial is easy to compute: $f(1) = 1$. In the recursive step, n is multiplied by the result of a recursive call to the factorial of $n - 1$.

TRY IT! Write the factorial function using recursion. Use your function to compute the factorial of 3.

```
function [out] = myRecFactorial(n)
% [out] = myRecFactorial(n)
% Recursive implementation of factorial function.
% author
% date

if n == 1
    % base case
    out = 1;
else
    % recursive step
    out = n*myRecFactorial(n-1);
end

end % end myRecFactorial


>> F = myRecFactorial(3)
F =
     6
```

WHAT IS HAPPENING? First recall that when MATLAB executes a function, it creates a workspace for the variables that are created in that function, and whenever a function calls another function, it will wait until that function returns an answer before continuing. For example, in the line » `sin(tan(x))`, `sin` must wait for `tan` to return an answer before it can be evaluated. Even though a recursive function makes calls to itself, the same rules apply.

1. User makes call to `myRecFactorial(3)`. A new workspace is opened to compute `myRecFactorial(3)`.

2. Input argument value, 3, is compared to 1. Since they are not equal, else statement is executed.
3. `3*myRecFactorial(2)` must be computed. A new workspace is opened to compute `myRecFactorial(2)`.
4. Input argument value, 2, is compared to 1. Since they are not equal, else statement is executed.
5. `2*myRecFactorial(1)` must be computed. A new workspace is opened to compute `myRecFactorial(1)`.
6. Input argument value, 1, is compared to 1. Since they are equal, if statement is executed.
7. Output variable out is assigned the value 1. `myRecFactorial(1)` terminates with output 1.
8. `2*myRecFactorial(1)` can be resolved to $2 * 1 = 2$. Output variable out is assigned the value 2. `myRecFactorial(2)` terminates with output 2.
9. `3*myRecFactorial(2)` can be resolved to $3 * 2 = 6$. Output variable out is assigned the value 6. `myRecFactorial(3)` terminates with output 3 to user.

The order of recursive calls can be depicted by a **recursion tree** shown in Figure 6.1 for `myRecFactorial(3)`. A recursion tree is a diagram of the function calls connected by numbered arrows to depict the order in which the calls were made.

Fibonacci numbers were originally developed to model the idealized population growth of rabbits. Since then, they have been found to be significant in any naturally occurring phenomena. The Fibonacci numbers can be generated using the following recursive formula. Note that the recursive step contains two recursive calls and that there are also two base cases (i.e., two cases that cause the recursion to stop).

$$F(n) = \begin{cases} 1 & \text{if } n = 1 \\ 1 & \text{if } n = 2 \\ F(n-1) + F(n-2) & \text{otherwise} \end{cases}$$

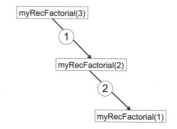

FIGURE 6.1

Recursion tree for `myRecFactorial(3)`.

TRY IT! Write a recursive function for computing the *n*-th Fibonacci number. Use your function to compute the first five Fibonacci numbers. Draw the associated recursion tree (see Figure 6.2).

```
function [F] = myRecFib(n)

if n == 1
    out = 1; % first base case
elseif n == 2
    out = 1; % second base case
else
    out = myRecFib(n-1) + myRecFib(n-2); % recursive call
end

end % end myRecFib
```

```
>> [F] = myRecFib(1)
F =
    1
>> [F] = myRecFib(2)
F =
    1
>> [F] = myRecFib(3)
F =
    2
>> [F] = myRecFib(4)
F =
    3
>> [F] = myRecFib(5)
F =
    5
```

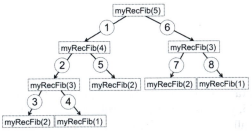

FIGURE 6.2

Recursion Tree for `myRecFib(5)`.

As an exercise, consider the following modification to myRecFib, where the results of each recursive call are displayed to the screen.

EXAMPLE: Modification for myRecFib. Can you determine the order in which the Fibonacci numbers will appear on the screen for myRecFib(5)?

```
function [F] = myRecFib(n)

if n == 1
    out = 1; % first base case
    disp(out)
elseif n == 2
    out = 1; % second base case
    disp(out)
else
    out = myRecFib(n-1) + myRecFib(n-2); % recursive call
    disp(out)
end

end % end myRecFib


[F] = myRecFib(5)
     1
     1
     2
     1
     3
     1
     1
     2
     5
F =
     5
```

Notice that the number of recursive calls becomes very large for even relatively small inputs for n. If you do not agree, try to draw the recursion tree for myRecFib(10). If you try your unmodified function for inputs around 25, you will notice significant computation times.

Every time a recursive call is made, MATLAB must create a new workspace for it, and all the current workspaces must be retained inside MATLAB's memory. If MATLAB runs out of memory, then it will crash (i.e., stop working and close down). To prevent this from happening, MATLAB has a **recursion limit** that restricts the number of workspaces that can be created by a recursive function. The default recursion limit is 500. You can increase this limit to N using the command set(0,'RecursionLimit',N). However doing so is not advisable unless absolutely necessary.

TRY IT! Run the unmodified myRecFib for n = 501.

```
>> myRecFib(501)
??? Maximum recursion limit of 500 reached. Use set(0,'RecursionLimit',N)
to change the limit.  Be aware that exceeding your available stack space can
crash MATLAB and/or your computer.

Error in ==> myRecFib
```

There is an iterative method of computing the n-th Fibonacci number that requires only one workspace.

EXAMPLE: Iterative implementation for computing Fibonacci numbers.

```
function [F] = myIterFib(n)

F = ones(1,n);
for i = 3:n
    F(i) = F(i-1) + F(i-2);
end

F = F(n);
end
```

TRY IT! Write a script to compute the 25-th Fibonacci number using myIterFib and myRecFib. Use the tic and toc functions to measure the run time of each function. Notice the large difference in running times.

```
%% Speed Test for Fibonacci Functions
clc
clear
close all

%% Iterative Implementation
tic
F = myIterFib(25)
tIter = toc

%% Recursive Implementation
tic
F = myRecFib(25)
tRec = toc
```

```
F =
        75025
tIter =
      0.0013
F =
        75025
tRec =
      1.5346
```

You can see in the previous example that the iterative version runs much faster than the recursive counterpart. In general, iterative functions are faster than recursive functions that perform the same task. So why use recursive functions at all? There are some solution methods that have a naturally recursive structure. In these cases it is usually very hard to write a counterpart using loops. The primary value of writing recursive functions is that they can usually be written much more compactly than iterative functions. The cost of the improved compactness is added running time.

The relationship between the input arguments and the running time is discussed in more detail in Chapter 7 on Complexity.

TIP! Try to write functions iteratively whenever it is convenient to do so. Your functions will run faster.

6.2 Divide and Conquer

Divide and conquer is a useful strategy for solving difficult problems. Using divide and conquer, difficult problems are solved from solutions to many similar easy problems. In this way, difficult problems are broken up so they are more manageable. In this section, we cover two classical examples of divide and conquer: the Towers of Hanoi Problem and the Quicksort algorithm.

Towers of Hanoi

The Towers of Hanoi problem consists of three vertical rods, or towers, and N disks of different sizes, each with a hole in the center so that the rod can slide through it. The disks are originally stacked on one of the towers in order of descending size (i.e., the largest disc is on the bottom). The goal of the problem is to move all the disks to a different rod while complying with the following three rules:

1. Only one disk can be moved at a time.
2. Only the disk at the top of a stack may be moved.
3. A disk may not be placed on top of a smaller disk.

Figure 6.3 shows the steps of the solution to the Towers of Hanoi problem with three disks.

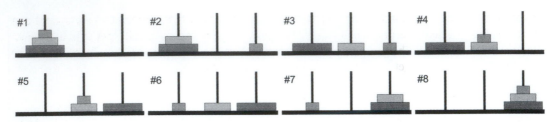

FIGURE 6.3

Illustration of the Towers of Hanoi: In eight steps, all disks are transported from pole 1 to pole 3, one at a time, by moving only the disk at the top of the current stack, and placing only smaller disks on top of larger disks.

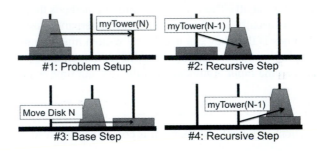

FIGURE 6.4

Breakdown of one iteration of the recursive solution of the Towers of Hanoi problem.

A legend goes that a group of Indian monks are in a monastery working to complete a Towers of Hanoi problem with 64 disks. When they complete the problem, the world will end. Fortunately, the number of moves required is $2^{64} - 1$ so even if they could move one disk per millisecond, it would take over 584 million years for them to finish.

The key to the Towers of Hanoi problem is breaking it down into smaller, easier-to-manage problems that we will refer to as **subproblems**. For this problem, it is relatively easy to see that moving a disk is easy (which has only three rules) but moving a tower is difficult (we cannot immediately see how to do it). So we will assign moving a stack of size N to several subproblems of moving a stack of size $N - 1$.

Consider a stack of N disks that we wish to move from Tower 1 to Tower 3, and let `myTower(N)` move a stack of size N to the desired tower (i.e., display the moves). How to write `myTower` may not immediately be clear. However, if we think about the problem in terms of subproblems, we can see that we need to move the top `N-1` disks to the middle tower, then the bottom disk to the right tower, and then the `N-1` disks on the middle tower to the right tower. `myTower` can display the instruction to move disk N, and then make recursive calls to `myTower(N-1)` to handle moving the smaller towers. The calls to `myTower(N-1)` make recursive calls to `myTower(N-2)` and so on. A breakdown of the three steps is depicted in Figure 6.4.

Following is a recursive solution to the Towers of Hanoi problem. Notice its compactness and simplicity. The code exactly reflects our intuition about the recursive nature of the solution: First we move a stack of size N-1 from the original tower 'from' to the alternative tower 'alt'. This is a difficult task, so instead we make a recursive call that will make subsequent recursive calls, but will, in the end, move the stack as desired. Then we move the bottom disk to the target tower 'to'. Finally, we move the stack of size N-1 to the target tower by making another recursive call.

```
function [] = myTowers(N, from, to, alt)
% myTowers(N, from, to, alt)
% Displays the moves required to move a tower of size N from the 'from' tower to the
% 'to' tower.
% from, to, and alt are uniquely either 1,2, or 3 referring to tower 1, tower 2,
% and tower 3.
%

if N ~= 0

    % recursive call that moves N-1 stack from starting tower to alternate tower
    myTowers(N-1, from, alt, to)

    % display to screen movement of bottom disk from starting tower to final tower
    disp(sprintf('Move disk %d from tower %d to tower %d.', N, from, to))

    % recursive call that moves N-1 stack from alternate tower to final tower
    myTowers(N-1, alt, to, from)
end

end % end myTowers
```

TRY IT! Use the function myTowers to solve the Towers of Hanoi Problem for N = 3. Verify that the solution is correct by inspection.

```
>> myTowers(3,1,3,2)
Move disk 1 from tower 1 to tower 3.
Move disk 2 from tower 1 to tower 2.
Move disk 1 from tower 3 to tower 2.
Move disk 3 from tower 1 to tower 3.
Move disk 1 from tower 2 to tower 1.
Move disk 2 from tower 2 to tower 3.
Move disk 1 from tower 1 to tower 3.
```

By using Divide and Conquer, we have solved the Towers of Hanoi problem by making recursive calls to slightly smaller Towers of Hanoi problems that, in turn, make recursive calls to yet smaller Towers of Hanoi problems. Together, the solutions form the solution to the whole problem. The actual work done by a single function call is actually quite small: two recursive calls and moving one disk. In other words, a function call does very little work (moving a disk), and then passes the rest of the work onto other calls, a skill you will probably find very useful throughout your engineering career.

QuickSort

An array of numbers, A, is **sorted** if the elements are arranged in ascending or descending order. Although there are many ways of sorting a list, quicksort is a divide-and-conquer approach that is a very fast algorithm for sorting using a single processor (there are faster algorithms for multiple processors).

The quicksort algorithm starts with the observation that sorting a list is hard, but comparison is easy. So instead of sorting a list, we separate the array by comparing to a **pivot**. At each recursive call to quicksort, the input array is divided into three parts: elements that are smaller than the pivot, elements that are equal to the pivot, and elements that are larger than the pivot. Then a recursive call to quicksort is made on the two subproblems: the array of elements smaller than the pivot and the array of elements larger than the pivot. Eventually the subproblems are small enough (i.e., array size of length 1 or 0) that sorting the list is trivial.

Consider the following recursive implementation of quicksort.

```
function [sorted] = myQuickSort(array)

if length(array) <= 1

    % array of length 1 is easiest to sort because it is already sorted
    sorted = array;

else

    % select pivot as the first element of the array
    pivot = array(1);

    % initialize arrays for bigger and smaller elements as well those equal to the pivot
    bigger = [];
    smaller = [];
    same = [];

    % loop through array and put elements into appropriate array
    for i = 1:length(array)

        if array(i) > pivot
            bigger = [bigger, array(i)];
        elseif array(i) < pivot
            smaller = [smaller, array(i)];
        else
            same = [same, array(i)];
        end

    end % end for i

    % make recursive calls to quicksort and concatenate solutions
    sorted = [myQuickSort(smaller), same, myQuickSort(bigger)];

end % end if

end % end myQuickSort
```

Similarly to Towers of Hanoi, we have broken up the problem of sorting (hard) into many comparisons (easy).

Summary

1. A recursive function is a function that calls itself.
2. Recursive functions are useful when problems have a hierarchical structure rather than an iterative structure.
3. Divide and Conquer is a powerful problem-solving strategy that can be used to solve difficult problems.

Vocabulary

base case	recursive step	sorted
Divide and Conquer	recursion tree	subproblem
recursive function		

Functions and Operators

Problems

⌊m 1. Write a function with header [S] = mySum(A) where A is a one-dimensional array, and S is the sum of all the elements of A. You can use recursion or iteration to solve the problem, but do not use MATLAB's function sum.

Test Cases:

```
>> S = mySum([1 2 3])
S =
      6
>> S = mySum(1:100)
S =
    5050
```

⌊m 2. Chebyshev polynomials are defined recursively. Chebyshev polynomials are separated into two kinds: first and second. Chebyshev polynomials of the first kind, $T_n(x)$, and of the second kind, $U_n(x)$, are defined by the following recurrence relations:

$$T_n(x) = \begin{cases} 1 & \text{if } n = 0 \\ x & \text{if } n = 1 \\ 2x\,T_{n-1}(x) - T_{n-2}(x) & \text{otherwise} \end{cases}$$

$$U_n(x) = \begin{cases} 1 & \text{if } n = 0 \\ 2x & \text{if } n = 1 \\ 2xU_{n-1}(x) - U_{n-2}(x) & \text{otherwise} \end{cases}$$

Write a function with header [y] = myChebyshevPoly1(n,x), where y is the n-th Chebyshev polynomial of the first kind evaluated at x. Be sure your function can take array inputs for x. You may assume that x is a row vector. The output variable, y, must be a row vector also.

Test Cases:

```
>> myChebyshevPoly1(0,1:5)
ans =
    1     1     1     1     1
>> myChebyshevPoly1(1,1:5)
ans =
    1     2     3     4     5
>> myChebyshevPoly1(3,1:5)
ans =
    1    26    99   244   485
```

Try plotting your Chebyshev polynomials of various orders for x = 0:.01:5 if you are interested in seeing what they looking like.

.m 3. The Ackermann function, A, is a quickly growing function that is defined by the recursive relationship:

$$A(m, n) = \begin{cases} n + 1 & \text{if } m = 0 \\ A(m - 1, 1) & \text{if } m > 0 \text{ and } n = 1 \\ A(m - 1, A(m, n - 1)) & \text{if } m > 0 \text{ and } n > 0 \end{cases}$$

Write a function with header [A] = myAckermann(m,n), where A is the Ackermann function computed for m and n.

Test Cases:

```
>> A = myAckermann(1,1)
A =
     3
>> A = myAckermann(1,2)
A =
     4
>> A = myAckermann(2,3)
A =
     9
>> A = myAckermann(3,3)
A =
    61
>> A = myAckermann(3,4)
A =
   125
```

myAckermann (4,4) is so large that it would be difficult to write down. Although the Acker-mann function does not have many practical uses, the inverse Ackermann function has several uses in robotic motion planning.

.m **4.** A function, $C(n, k)$, which computes the number of different ways of uniquely choosing k objects from n without repetition, is commonly used in many statistics applications. For exam-ple, how many three-flavored ice cream sundaes are there if there are 10 icecream flavors? To solve this problem we would have to compute $C(10, 3)$, the number of ways of choosing three unique icecream flavors from 10. The function C is commonly called "n choose k." You may assume that n and k are 1×1 integer doubles.

If $n = k$, then clearly $C(n, k) = 1$ because there is only way to choose n objects from n objects. If $k = 1$, then $C(n, k) = n$ because choosing each of the n objects is a way of choosing one object from n. For all other cases, $C(n, k) = C(n - 1, k) + C(n - 1, k - 1)$. Can you see why? Write a function with header [N] = myNChooseK (n, k) that computes the number of times k objects can be uniquely chosen from n objects without repetition.

Test Cases:

```
>> N = myNChooseK (10,1)
N =

            10
>> N = myNChooseK (10,10)
N =

            1
>> N = myNChooseK (10,3)
N =

         120
```

.m **5.** In purchases paid in cash, the seller must return money that was overpaid. This is commonly referred to as "giving change." The bills and coins required to properly give change can be defined by a recursive relationship. If the amount paid is more than $100 more than the cost, then return a hundred-dollar bill along with the result of a recursive call to the change func-tion with $100 subtracted from the amount paid. If the amount paid is more than $50 over the cost of the item, then return a fifty-dollar bill, along with the result of a recursive call to the change function with $50 subtracted. Similar clauses can be given for every denom-ination of US currency. The denominations of US currency, in dollars, is 100, 50, 20, 10, 5, 1, 0.25, 0.10, 0.05, and 0.01. For this problem we will ignore the two-dollar bill, which is not in common circulation. You may assume that cost and paid are scalars, and that $paid >= cost$. The output variable, change, must be a column vector as shown in the test case.

Use recursion to program a function with header [change] = myChange (cost, paid) where cost is the cost of the item, paid is the amount paid, and change is an array con-taining the list of bills and coins that should be returned to the seller. Note: Watch out for the base case!

Test Cases:

```
>> format ban
>> C = myChange(27.57, 100)
C =
            50.00
            20.00
             1.00
             1.00
             0.25
             0.10
             0.05
             0.01
             0.01
             0.01
```

.m 6. The golden ratio, ϕ, is the limit of $\frac{F(n+1)}{F(n)}$ as n goes to infinity and $F(n)$ is the n-th Fibonacci number, which can be shown to be exactly $\frac{1+\sqrt{5}}{2}$ and is approximately 1.62. We say that $G(n) = \frac{F(n+1)}{F(n)}$ is the n-th approximation of the golden ratio, and $G(1) = 1$.

It can be shown that ϕ is also the limit of the continued fraction:

$$\varphi = 1 + \cfrac{1}{1 + \cfrac{1}{1 + \cfrac{1}{1 + \cfrac{1}{1 + \cdots}}}}.$$

Write a recursive function with header [G] = myGoldenRatio(n), where G is the n-th approximation of the golden ratio according to the continued fraction recursive relationship. You should use the continued fraction approximation for the Golden ratio, not the $G(n) = F(n+1)/F(n)$ definition. However for both definitions, $G(1) = 1$.

Test Cases:

```
>> format long
>> G = myGoldenRatio(10)
G =
    1.618181818181818
>> (1 + sqrt(5))/2
ans =
    1.618033988749895
```

Studies have shown that rectangles with aspect ratio (i.e., length divided by width) close to the golden ratio are more pleasing than rectangles that do not. What is the aspect ratio of many wide-screen TV's and movie screens?

.m 7. The greatest common divisor of two integers a and b is the largest integer that divides both numbers without remainder, and the function to compute it is denoted by GCD(a,b). The

GCD function can be written recursively. If b equals 0, then a is the greatest common divisor. Otherwise, GCD(a,b) = GCD(b, rem(a,b)) where rem(a,b) is the remainder of a divided by b. Assume that a and b are 1×1 integer doubles.

Write a recursive function with header [gcd] = myGCD(a,b) that computes the greatest common divisor of a and b. Assume that a and b are 1×1 integer doubles.

Test Cases:

```
>> gcd = myGCD(10,4)
gcd =
    2
>> gcd = myGCD(33,121)
gcd =
    11
>> gcd = myGCD(18,1)
gcd =
    1
```

8. Pascal's triangle is an arrangement of numbers such that each row is equivalent to the coefficients of the binomial expansion of $(x + y)^{(p-1)}$, where p is some positive integer more than or equal to 1. For example, $(x + y)^2 = 1x^2 + 2xy + 1y^2$ so the third row of Pascal's triangle is 1 2 1. Let R_m represent the m-th row of Pascal's triangle, and $R_m(n)$ be the n-th element of the row. By definition, R_m has m elements, and $R_m(1) = R_m(n) = 1$. The remaining elements are computed by the following recursive relationship: $R_m(i) = R_{m-1}(i-1) + R_{m-1}(i)$ for $i = 2, \ldots, m-1$. The first few rows of Pascal's triangle are depicted in the following figure. You may assume that m is a strictly positive integer. The output variable, row, must be a row vector.

```
            1
          1   1
        1   2   1
      1   3   3   1
    1   4   6   4   1
  1   5  10  10   5   1
```

Write a function with header [row] = myPascalRow(m) where row is the m-th row of the Pascal triangle. You may assume that m is a strictly positive integer.

Test Cases:

```
>> R = myPascalRow(1)
R =
    1
>> R = myPascalRow(2)
R =
    1    1
>> R = myPascalRow(3)
R =
    1    2    1
```

```
>> R = myPascalRow(4)
R =
     1    3    3    1
>> R = myPascalRow(5)
R =
     1    4    6    4    1
>>
```

m 9. Consider a $n \times n$ matrix of the following form:

$$A = \begin{bmatrix} 1 & 1 & 1 & 1 & 1 \\ 1 & 0 & 0 & 0 & 0 \\ 1 & 0 & 1 & 1 & 0 \\ 1 & 0 & 0 & 1 & 0 \\ 1 & 1 & 1 & 1 & 0 \end{bmatrix}$$

where the ones form a right spiral. Write a function with header `[A] = mySpiralOnes(n)` that produces an $n \times n$ matrix of the given form. Take care that the recursive steps are in the correct order (i.e., the ones go right, then down, then left, then up, then right, etc.).

Test Cases:

```
>> mySpiralOnes(1)
ans =
     1
>> mySpiralOnes(2)
ans =
     1    1
     0    1
>> mySpiralOnes(3)
ans =
     0    1    1
     0    0    1
     1    1    1
>> mySpiralOnes(4)
ans =
     1    0    0    0
     1    0    1    1
     1    0    0    1
     1    1    1    1
>> mySpiralOnes(5)
ans =
     1    1    1    1    1
     1    0    0    0    0
     1    0    1    1    0
     1    0    0    1    0
     1    1    1    1    0
```

m 10. Rewrite `mySpiralOnes` without using recursion.

● **11.** Write a line of code that produces the following error:

```
??? Maximum recursion limit of 500 reached. Use set(0,'RecursionLimit',N)
to change the limit.  Be aware that exceeding your available stack space
can crash MATLAB and/or your computer.
```

12. Draw the recursion tree for `myTowers(4)`.

m 13. Rewrite the Towers of Hanoi function in this chapter without using recursion.

14. Draw the recursion tree for `myQuickSort([5 4 6 2 9 1 7 3])`.

m 15. Rewrite the quicksort function in this chapter without using recursion.

Complexity

CHAPTER OUTLINE

Motivation

Once you have programmed a solution to a problem, an important question is "How long is my program going to run?" Clearly the answer to this question depends on many factors, such as the computer memory, the computer speed, and the size of the problem. For example, if your function sums every element of a very large array, the time to complete the task will depend on whether your computer can hold the entire array in its memory at once, how fast your computer can do additions, and the size of the array.

The effort required to run a program to completion is the notion of "complexity," and it is the topic of this chapter. By the end of this chapter, you should be able to estimate the complexity of simple programs and identify poor complexity properties when you see them.

7.1 Complexity and Big-O Notation

The **complexity** of a function is the relationship between the size of the input and the difficulty of running the function to completion. The size of the input is usually denoted by n. In computer science, this is usually taken to be the number of bits required to describe the problem. However, n usually describes something more tangible, such as the length of an array. The difficulty of a problem can be measured in several ways. It can be measured by the number of **bit operations** needed for the function to finish, which means the number of times a 1 must be turned to a 0 and vice versa, as well as a few other simple things that computers can do. It is usually more suitable to describe the difficulty of the problem in terms of **basic operations**: additions, subtractions, multiplications, divisions, assignments, and function calls. Although each basic operation takes different amounts of time, the number of basic operations needed to complete a function is sufficiently related to the running time to be useful, and it is much easier to count.

An Introduction to MATLAB® Programming and Numerical Methods. http://dx.doi.org/10.1016/B978-0-12-420228-3.00007-5

TRY IT! Count the number of basic operations, in terms of n, required for the following function to terminate.

```matlab
function [out] = f(n)
% author
% date

out = 0;
for i = 1:n
    for j = 1:n
        out = out + i*j;
    end
end
end % end f
```

additions: n^2, subtractions: 0, multiplications: n^2, divisions: 0, assignments: $2n^2 + 1$, function calls: 0, total: $4n^2 + 1$.

The number of assignments is $2n^2 + n + 1$ because the line out = out + i*j is evaluated n^2 times, j is assigned n^2, i is assigned n times, and the line out = 0 is assigned once. So, the complexity of the function f can be described as $4n^2 + n + 1$.

A common notation for complexity is called **Big-O notation**. Big-O notation establishes the relationship in the *growth* of the number of basic operations with respect to the size of the input as the input size becomes very large. As n gets large, the highest power dominates; therefore, only the highest power term is included in Big-O notation. Additionally, coefficients are not required to characterize growth, and so coefficients are also dropped. In the previous example, we counted $4n^2 + n + 1$ basic operations to complete the function. In Big-O notation we would say that the function is $O(n^2)$ (pronounced "O of n-squared"). We say that any algorithm with complexity $O(n^c)$ where c is some constant with respect to n is **polynomial time**.

TRY IT! Determine the complexity of the iterative Fibonacci function in Big-O notation.

```matlab
function [F] = myFibIter(n)

F = [1 1];

for i = 3:n
    F(i) = F(i-1) + F(i-2);
end

F = F(end);

end % end myFibIter
```

Since the only lines of code that take more time as n grows are those in the for-loop, we can restrict our attention to the for-loop and the code block within it. The code within the for-loop does not grow with respect to n (i.e., it is constant). Therefore, the number of basic operations is Cn where C is some constant representing the number of basic operations that occur in the for-loop, and these C operations run n times. This gives a complexity of $O(n)$ for `myFibIter`.

Assessing the exact complexity of a function can be difficult. In these cases, it might be sufficient to give an upper bound or even an approximation of the complexity.

TRY IT! Give an upper bound on the complexity of the recursive implementation of Fibonacci. Do you think it is a good approximation of the upper bound? Do you think that recursive Fibonacci could possibly be polynomial time?

```
function [F] = myFibRec(n)

if n <= 2
    F = 1;
else
    F = myFibRec(n−1) + myFibRec(n−2);
end

end % end myFibRec
```

As n gets large, we can say that the vast majority of function calls make two other function calls: one addition and one assignment to the output. The addition and assignment do not grow with n per function call, so we can ignore them in Big-O notation. However, the number of function calls grows approximately by 2^n, and so the complexity of `myFibRec` is upper bound by $O(2^n)$.

There is on-going debate whether or not $O(2^n)$ is a good approximation for the Fibonacci function.

Since the number of recursive calls grows exponentially with n, there is no way the recursive fibonacci function could be polynomial. That is, for any c, there is an n such that `myFibRec` takes more than $O(n^c)$ basic operations to complete. Any function that is $O(c^n)$ for some constant c is said to be **exponential time**.

TRY IT! What is the complexity of the following function in Big-O notation?

```
function [out] = myDivideByTwo(n)

out = 0;
while n > 1
    n = n/2;
    out = out + 1;
end

end % end myDivideByTwo
```

Again, only the while-loop runs longer for larger n so we can restrict our attention there. Within the while-loop, there are two assignments: one division and one addition, which are both constant time with respect to n. So the complexity depends only on how many times the while-loop runs. The while-loop cuts n in half in every iteration until n is less than 1. So the number of iterations, I, is the solution to the equation $\frac{n}{2^I} = 1$. With some manipulation, this solves to $I = \log n$, so the complexity of myDivideByTwo is $O(\log n)$. It does not matter what the base of the log is because, recalling log rules, all logs are a scalar multiple of each other. Any function with complexity $O(\log n)$. is said to be **log time**.

7.2 Complexity Matters

So why does complexity matter? Assume you have an algorithm that runs in exponential time, say $O(2^n)$, and let N be the largest problem you can solve with this algorithm using the computational resources you have, denoted by R. R could be the amount of time you are willing to wait for the function to finish, or R could be the number of basic operations you watch the computer execute before you get sick of waiting. Using the same algorithm, how large of a problem can you solve given a new computer that is twice as fast?

If we establish $R = 2^N$, using our old computer, with our new computer we have $2R$ computational resources; therefore, we want to find N' such that $2R = 2^{N'}$. With some substitution, we can arrive at $2 \times 2^N = 2^{N'} \rightarrow 2^{N+1} = 2^{N'} \rightarrow N' = N + 1$. So with an exponential time algorithm, doubling your computational resources will allow you to solve a problem one unit larger than you could with your old computer. This is a very small difference. In fact as N gets large, the relative improvement goes to 0.

With a polynomial time algorithm, you can do much better. This time let's assume that $R = N^c$, where c is some constant larger than one. Then $2R = N'^c$, which using similar substitutions as before gets you to $N' = 2^{1/c}N$. So with a polynomial time algorithm with power c, you can solve a problem $\sqrt[c]{2}$ larger than you could with your old computer. When c is small, say less than 5, this is a much bigger difference than with the exponential algorithm.

Finally, let us consider a log time algorithm. Let $R = \log N$. Then $2R = \log N'$, and again with some substitution we obtain $N' = N^2$. So with the double resources, we can square the size of the problem we can solve!

The moral of the story is that exponential time algorithms do not scale well. That is, as you increase the size of the input, you will soon find that the function takes longer (much longer) than you are willing to wait. For one final example, myFibRec(100) would take on the order 2^{100} basic operations to perform. If your computer could do 100 trillion basic operations per second (far faster than the fastest computer on earth), it would take your computer about 400 million years to complete. However, myFibIter(100) would take less than 1 nanosecond.

There is both an exponential time algorithm (recursion) and a polynomial time algorithm (iteration) for computing Fibonacci numbers. Given a choice, we would clearly pick the polynomial time algorithm. However, there is a class of problems for which no one has ever discovered a polynomial time algorithm. In other words, there are only exponential time algorithms known for them. These problems are known as NP-Complete, and there is ongoing investigation as to whether polynomial time algorithms exist for these problems. Examples of NP-Complete problems include the Traveling Salesman, Set Cover, and Set Packing problems. Although theoretical in construction, solutions to these problems have numerous applications in logistics and operations research. In fact, some encryption algorithms that keep web and bank applications secure rely on the NP-Complete-ness of breaking them. A further discussion of NP-Complete problems and the theory of complexity is beyond the scope of this book but these problems are very interesting and important to many engineering applications.

7.3 **The Profiler**

Even if it does not change the Big-O complexity of a program, many programmers will spend long hours to make their code run twice as fast or to gain even smaller improvements. The **MATLAB Profiler** is a useful tool for improving the performance of your programs. You can open the Profiler under Desktop → Profiler.

TRY IT! Consider the following script, ProfilerTest.m, that sums random numbers over and over again. Run ProfilerTest.m in the Profiler.

```
%% Profiler Test
clc
clear
close all

% define number of iterations and size of random list.
N = 1000;
n = 10000;

% run through iterations
for i = 1:N

    % create random numbers
    A = rand(1,n);

    % do sum
    S = 0;
    for j = 1:n

        S = S + A(j);

    end
end
```

You can run your code in the Profiler by typing into the box next to "Run this code" as if it were at the command window. You can execute the code by pressing Enter (PC) or Return (Mac).

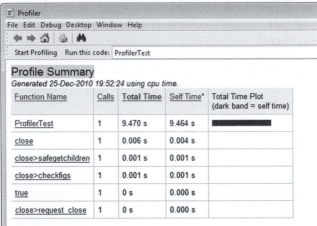

FIGURE 7.1

Profiler overall results.

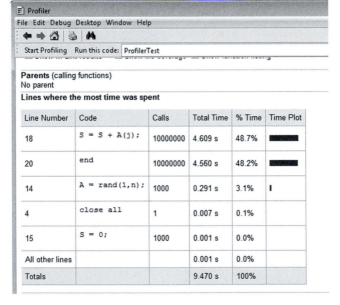

FIGURE 7.2

Profiler results for `ProfilerTest.m`.

Click on the `ProfilerTest` link (see Figure 7.1; first in the list) and you will be taken to a breakdown of the time spent on each line in `ProfilerTest.m`. You can see (Figure 7.2) that most of the time is spent on the line `S = S + A(j)` and for driving the for-loop (for-loop end statement).

You can almost always improve the performance of your program by using a MATLAB function that accomplishes the same thing. In this case, the MATLAB function `sum` does the same task as the inner for-loop in `ProfilerTest.m`.

TIP! Use MATLAB built-in functions whenever possible (i.e., sum, sqrt, find, etc.).

TRY IT! Replace the inner for-loop in `ProfilerTest.m` with the `sum` function. Run the modified script in the Profiler.

```
%% Profiler Test (Modified)
clc
clear
close all

% define number of iterations and size of random list.
N = 1000;
n = 10000;

% run through iterations
for i = 1:N

    % create random numbers
    A = rand(1,n);

    % do sum
    S = sum(A);

end
```

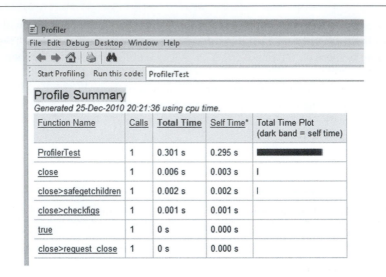

FIGURE 7.3

Improved profiler results for `ProfilerTest.m`.

As you can see in Figure 7.3, the overall performance of the code has been significantly improved.

Usually when code takes longer to run than you would like, there is a **bottleneck** where much of the time is being spent. That is, there is a line of code that is taking much longer to execute than the other lines in the program. Addressing the bottleneck in a program will usually lead to the biggest improvement in performance, even if there are other areas of your code that are more easily improved.

> **TIP!** Start at the bottleneck when improving the performance of code.

Summary

1. The complexity of an algorithm is the relationship between the size of the input problem and the time it takes the algorithm to terminate.
2. Big-O notation is a standard method of classifying algorithmic complexity in a way that is computer- and operating-system-independent.
3. Algorithms with log complexity are faster than algorithms with polynomial complexity, which are faster than algorithms with exponential complexity.
4. The Profiler is a useful MATLAB tool for determining where your code is running slowly so that you can improve its performance.

Vocabulary

basic operation	bottleneck	polynomial time
Big-O notation	complexity	Profiler
bit operation	exponential time	

Functions and Operators

bin2dec   dec2bin

eps

Problems

1. How would you define the size of the following tasks?

    1. Solving a jigsaw puzzle.
    2. Passing a handout to a class.
    3. Walking to class.
    4. Finding a name in dictionary.

2. For the tasks given in the previous problem, what would you say is the Big-O complexity of the tasks in terms of the size definitions you gave?

3. You may be surprised to know that there is a log time algorithm for finding a word in an n-word dictionary. Instead of starting at the beginning of the list, you go to the middle. If this is the word you are looking for then you are done. If the word comes after the word you are looking for, then look halfway between the current word and the end. If it is before the word you are looking for, then look halfway between the first word and the current word. Keep repeating this process until you find the word. This algorithm is known as a binary search, and it is log time because the search space is cut in half at each iteration, and therefore, requires at most $\log_2(n)$ iterations to find the word. Hence the increase in run time is only log in the length of the list. There is a way to look up a word in $O(1)$ or constant time. This means that no matter how long the list is, it takes the same amount of time! Can you think of how this is done? Hint: Research hash functions.

4. What is the complexity of the algorithms that compute the following recursive relationships? Classify the following algorithms as log time, polynomial time, or exponential time in terms of n given that the implementation is (a) recursive, and (b) iterative.

    Tribonacci, $T(n)$:

    $$T(n) = T(n-1) + T(n-2) + T(n-3)$$
    $$T(1) = T(2) = T(3) = 1.$$

Timmynacci, $t(n)$:

$$t(n) = t(n/2) + t(n/4)$$
$$t(n) = 1 \text{ if } n < 1.$$

5. What is the Big-O complexity of the Towers of Hanoi problem given in Chapter 7? Is the complexity an upper bound or is it exact?

6. What is the Big-O complexity of the quicksort algorithm?

7. Run the following two iterative implementations of finding Fibonacci numbers in the MATLAB Profiler. The first implementation preallocates memory to an array that stores all the Fibonacci numbers. The second implementation expands the array at each iteration of the for-loop.

```
function [F] = myFibIter1(n)

F = zeros(1,n);
F(1,[1,2]) = 1;
for i = 3:n
    F(i) = F(i−1) + F(i−2);
end

F = F(end);

end
```

```
function [F] = myFibIter2(n)

F = [1 1];
for i = 3:n
    F = [F, F(end) + F(end−1)];
end

F = F(end)

end
```

Which function runs faster for $n = 100$?

Representation of Numbers

CHAPTER OUTLINE

Motivation

There are many ways of representing or writing numbers. For example, decimal numbers, Roman numerals, scientific notation, and even tally marks are all ways of representing numbers (Figure 8.1).

Mathematics has infinite precision when describing numbers. For example, $\sqrt{2}$ is *precisely* the number such that when you square it, the result is 2, the exact decimal representation of which requires an infinite number of digits. This presents a significant problem for computers, since they only have a limited amount of space to store numbers, and they cannot understand abstract notations of representing numbers more compactly.

In this chapter, you will learn about different representation of numbers and how they are useful for computers. By the end of the chapter, you should know and understand some representations of numbers that are used in computing, how to convert between them and decimal numbers, and their primary advantages and disadvantages.

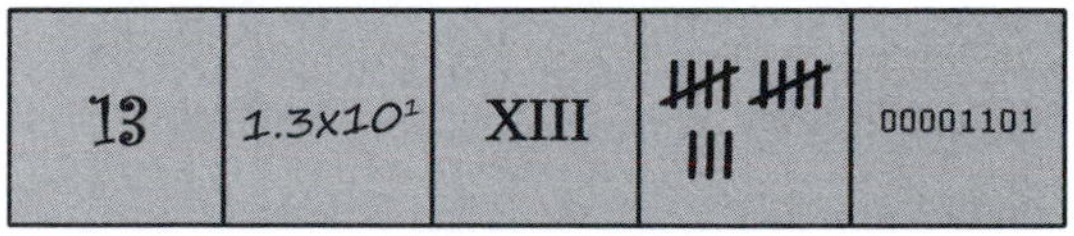

FIGURE 8.1

Various representations of the number 13. (left to right) Base10, scientific notation, Roman numerals, tally marks, and binary.

An Introduction to MATLAB® Programming and Numerical Methods. http://dx.doi.org/10.1016/B978-0-12-420228-3.00008-7

8.1 Base-N and Binary

The **decimal system** is a way of representing numbers that you are familiar with from elementary school. In the decimal system, a number is represented by a list of digits from 0 to 9, where each digit represents the coefficient for a power of 10.

EXAMPLE: Show the decimal expansion for 147.3.

$$147.3 = 1 \cdot 10^2 + 4 \cdot 10^1 + 7 \cdot 10^0 + 3 \cdot 10^{-1}.$$

Since each digit is associated with a power of 10, the decimal system is also known as **base10** because it is based on 10 digits (0 to 9). However, there is nothing special about base10 numbers except perhaps that you are more accustomed to using them. For example, in base3 we have the digits 0, 1, and 2 and the number $121(\text{base } 3) = 1 \cdot 3^2 + 2 \cdot 3^1 + 1 \cdot 3^0 = 9 + 6 + 1 = 16(\text{base } 10)$.

For the purposes of this chapter, it is useful to denote which representation a number is in. Therefore in this chapter, every number will be followed by its representation in parentheses (e.g., 11 (base10) means 11 in base10) unless the context is clear.

A very important representation of numbers for computers is base2 or **binary** numbers. In binary, the only available digits are 0 and 1, and each digit is the coefficient of a power of 2. Digits in a binary number are also known as a **bit**. Note that binary numbers are still numbers, and so addition and multiplication are defined on them exactly as you learned in grade school.

TRY IT! Convert the number 11 (base10) into binary.
$$11(\text{base10}) = 8 + 2 + 1 = 1 \cdot 2^3 + 0 \cdot 2^2 + 1 \cdot 2^1 + 1 \cdot 2^0 = 1011(\text{base2})$$

TRY IT! Convert 37 (base10) and 17 (base10) to binary. Add and multiply the resulting numbers in binary. Verify that the result is correct in base10.
Convert to binary: $37 \ (\text{base10}) = 32 + 4 + 1 = 1 \cdot 2^5 + 0 \cdot 2^4 + 0 \cdot 2^3 + 1 \cdot 2^2 + 0 \cdot 2^1 + 1 \cdot 2^0 = 100101 \ (\text{base2})$ $17 \ (\text{base10}) = 16 + 1 = 1 \cdot 2^4 + 0 \cdot 2^3 + 0 \cdot 2^2 + 0 \cdot 2^1 + 1 \cdot 2^0 = 10001 \ (\text{base2})$

Get results of addition and multiplication in decimal:
$37 + 17 = 54$
$37 \times 17 = 629$

Do addition in binary:

```
       1
  100101
+  10001
─────────
        0

       1
  100101
+  10001
─────────
       10

       1
  100101
+  10001
─────────
      110

       1
  100101
+  10001
─────────
     0110

       1
  100101
+  10001
─────────
    10110

       1
  100101
+  10001
─────────
   110110  = 32 + 16 + 4 + 2 + 0 = 54 (base10)
```

Do multiplication in binary:

```
    100101
  x  10001
  ─────────


    100101
  x  10001
  ─────────
    100101


    100101
  x  10001
  ─────────
    100101
         0
        00
       000
+1001010000
─────────────
 1001110101 = 512 + 64 + 32 + 16 + 4 + 1 = 629 (base10)
```

Binary numbers are useful for computers because arithmetic operations on the digits 0 and 1 can be represented using AND, OR, and NOT, which computers can do extremely fast.

Unlike humans that can abstract numbers to arbitrarily large values, computers have a fixed number of bits that they are capable of storing at one time. For example, a 32-bit computer can represent and process 32-digit binary numbers and no more. If all 32-bits are used to represent positive integer binary numbers, then this means that there are $\sum_{n=0}^{31} 2^n = 4,294,967,296$ numbers the computer can represent. This is not very many numbers at all and would be completely insufficient to do any useful arithmetic on. For example, you could not compute the perfectly reasonable sum $0.5 + 1.25$ using this representation because all the bits are dedicated to only positive integers.

8.2 Floating Point Numbers

The number of bits is usually fixed for any given computer. Using binary representation gives us an insufficient range and precision of numbers to do relevant engineering calculations. To achieve the range of values needed with the same number of bits, we use **floating point** numbers or **float** for short. There are two kinds of floats: **single** precision (32 bits) and **double** precision (64 bits). MATLAB's standard arithmetic uses double precision, but this section uses single precision to illustrate the concept of floating point numbers (to save space). Instead of utilizing each bit as the coefficient of a power of 2, floats allocate bits to three different parts: the **sign indicator**, s, which says whether a number is positive or negative; **characteristic** or **exponent**, e, which is the power of 2; and the **fraction**, f, which is the coefficient of the exponent. In the **IEEE754** standard for single precision, 1 bit is allocated to the sign indicator, 8 bits are allocated to the exponent, and 23 bits are allocated to the fraction. With 8 bits allocated to the exponent, this makes 256 values that this number can take. Since we want to be able to make very precise numbers, we want some of these values to represent negative exponents (i.e., to allow numbers that are between 0 and 1 (base 10)). To accomplish this, 127 is subtracted from the exponent to normalize it. The value subtracted from the exponent is commonly referred to as the **bias**. The fraction is a number between 1 and 2. In binary, this means that the leading term will always be 1, and, therefore, it is a waste of bits to store it. To save space, the leading 1 is dropped. A float can then be represented as

$$n = -1^s 2^{e-127}(1 + f).$$

TRY IT! What is the number 1 10000010 10000000000000000000000 (IEEE754) in base 10?

The exponent in decimal is $1 \cdot 2^7 + 1 \cdot 2^1 - 127 = 3$. The fraction is $1 \cdot \frac{1}{2^1} + 0 \cdot \frac{1}{2^2} + \ldots = 0.5$. Therefore $n = -1^1 \cdot 2^3 \cdot (1 + 0.5) = -12$ (base 10).

TRY IT! What is 15 (base 10) in IEEE754? What is the largest number smaller than 15? What is the smallest number larger than 15?

Since the number is positive, $s = 0$. The largest power of two that is smaller than 15 is 8, so the exponent is 3, making the characteristic $3 + 127 = 130$ (base 10) $= 10000010$ (base 2). Then the fraction is $15/8 - 1 = 0.875$ (base10) $= 1 \cdot \frac{1}{2^1} + 1 \cdot \frac{1}{2^2} 1 \cdot \frac{1}{2^3} = .11100000000000000000000$ (base2). When put together this produces the following conversion:

15 (base10) = 0 1000010 11100000000000000000000 (IEEE754).

The next smallest number is 0 1000010 11011111111111111111111 = 14.999999046325684.

The next largest number is 0 1000010 11100000000000000000001 = 15.000000953674316.

Therefore, the IEEE754 number 0 1000010 11100000000000000000000 not only represents the number 15, but also all the real numbers halfway between its immediate neighbors, or more specifically, the interval [14.999999523162842 15.000000476837158]. So any computation that has a result within this interval will be assigned 15.

We call the distance from one number to the next the **gap**. Because the fraction is multiplied by 2^{c-127}, the gap grows as the number represented grows. The gap at a given number can be computed using the function `eps`.

TRY IT! Use the `eps` function to determine the gap at 1e9. Verify that adding a number to 1e9 that is less than half the gap at 1e9 results in the same number.

```
>> eps(1e9)
ans =
    1.192092895507813e-007

>> 1e9 == (1e9+eps(1e9)/3)
ans =
    1
```

There are special cases for the value of a single-precision floating point number when $e = 0$ (i.e., $e = 00000000$ (base2)) and when $e = 255$ (i.e., $e = 11111111_2$ (base2)). When the exponent is 0, the leading 1 in the fraction takes the value 0 instead. The result is a **subnormal number**, which is computed by $n = -1^s 2^{-126}(0 + f)$. When the exponent is 255 and f is nonzero, then the result is "Not a Number," which MATLAB displays as NaN. This means that the number is undefined. When the exponent is 255, then $f = 0$ and $s = 0$, and the result is positive infinity, which MATLAB displays as Inf. When the exponent is 255, then $f = 0$, and $s = 1$, and the result is minus infinity, which MATLAB displays as -Inf. There are similar special rules for double-precision floating point numbers.

TRY IT! Compute the base10 value for 0 11111110 11111111111111111111111 (IEEE754), the largest defined number for 32 bits, and for 0 00000001 00000000000000000000001 (IEEE754), the smallest. Note the that the exponent is, respectively, $e = 254$ and $e = 1$ to comply with the previously stated rules. Verify that MATLAB agrees with these calculations using `realmax('single')` and `realmin('single')`.

```
>> format long

>> largest = (2^(254-127))*((1 + sum(2.^[-1:-1:-23]))))
largest =
     3.402823466385289e+38

>> smallest = (2^(1-127)) * (1+2^-23)
smallest =

     1.175494490952134e-38

>> realmax('single')
ans =
   3.4028235e+38

>> realmin('single')
ans =
   1.1754944e-38
```

Numbers that are larger than the largest representable floating point number result in **overflow**, and MATLAB handles this case by assigning the result to `Inf`. Numbers that are smaller than the smallest subnormal number result in **underflow**, and MATLAB handles this case by assigning the result to 0.

TRY IT! Recall that MATLAB uses double-precision floating point numbers as a default. Double-precision uses 1 sign bit, 11 exponent bits and a bias of 1023, and 52 fraction bits. Show that adding the maximum double-precision number with the gap at this number results in overflow and that MATLAB assigns this overflow number to `Inf`. Show that adding the maximum double-precision number to one-third the gap does not result in overflow.

```
>> format long
>> realmax('double')+eps(realmax('double'))
ans =
   Inf

>> realmax('double')+eps(realmax('double'))/3
ans =
     1.797693134862316e+308
```

TRY IT! The smallest subnormal number in double-precision has $s = 0$, $e = 00000000000$, and $f = 0000000000000000000000000000000000000000000000000001$. Using the special rules for subnormal numbers, this results in the subnormal number $-1^0 2^{1-1023} 2^{-52} = 2^{-1074}$. Show that 2^{-1075} underflows to 0 and that the result cannot be distinguished from 0. Show that 2^{-1074} does not.

```
>> 2^-1075
ans =
      0

>> 2^-1075 == 0
ans =
      1

>> 2^-1074
ans =
      4.940656458412465e-324
```

So, what have we gained by using IEEE754 versus binary? Using 32 bits gives us about 4 billion (2^{32}) numbers. Since the number of bits does not change between binary and IEEE754, IEEE754 must also give us about 4 billion numbers. In binary, numbers have a constant spacing between them. As a result, you cannot have both range (i.e., large distance between minimum and maximum representable numbers) and precision (i.e., small spacing between numbers). Controlling these parameters would depend on where you put the decimal point in your number. IEEE754 overcomes this limitation by using very high precision at small numbers and very low precision at large numbers. This limitation is usually acceptable because the gap at large numbers is still small *relative* to the size of the number itself. Therefore, even if the gap is millions large, it is irrelevant to normal calculations if the number under consideration is in the trillions or higher.

Summary

1. Numbers have many representations, and each representation has advantages and disadvantages.
2. Computers must represent numbers using a finite number of digits (bits).
3. Binary and IEEE754 are finite representations of numbers used by computers.

Vocabulary

base10	exponent	sign indicator (IEEE754)
bias	float	single
binary	floating point	subnormal number
bit	fraction (IEEE754)	underflow
characteristic (IEEE754)	gap	
double	IEEE754	
decimal system	overflow	

Functions and Operators

```
bin2dec
dec2bin
double
eps
```

Problems

⌐m 1. Write a function with header [d] = myBin2Dec(b) where b is a binary number represented by a one-dimensional array of ones and zeros. The last element of b represents the
coefficient of 2^0, the second-to-last element of b represents the coefficient of 2^1, and so on.
The output variable, d, should be the decimal representation of b.

Test Cases:

```
>> d = myBin2Dec([1 1 1])
d =
        7
>> d = myBin2Dec([0 0 0 0 0])
d =
        0
>> d = myBin2Dec([1 0 1 0 1 0 1])
d =
       85
>> d = myBin2Dec(ones(1,25))
d =
      33554431
```

⌐m 2. Write a function with header [b] = myDec2Bin(d), where d is a positive integer in
decimal, and b is the binary representation of d. The output b must be a row vector of ones
and zeros, and the leading term must be a 1 unless the decimal input value is 0.

Test Cases:

128   8 Representation of Numbers

```
>> b = myDec2Bin(0)
b =
        0
>> b = myDec2Bin(23)
b =
        1    0    1    1    1
>> b = myDec2Bin(2097)
b =
    1    0    0    0    0    0    1    1    0    0    0    1
```

>> **3.** Use the two functions you wrote in problems 1 and 2 to compute d = myBin2Dec (myDec2 Bin (12654)). Do you get the same number?

m **4.** Write a function with header [b] = myBinAdder (b1,b2), where b1, b2, and b are binary numbers represented as in problem 1. The output variable should be computed as b = b1 + b2. Do not use your functions from problems 1 and 2 to write this function (i.e., do not convert b1 and b2 to decimal; add them, and then convert the result back to binary). This function should be able to accept inputs b1 and b2 of any length (i.e., very long binary numbers), and b1 and b2 may not necessarily be the same length.

Test Cases:

```
>> b = myBinAdder ([1 1 1 1 1],[1])
b =
     1     0     0     0     0     0
>> b = myBinAdder ([1 1 1 1 1],[1 0 1 0 1 0 0])
b =
     1     1     1     0     0     1     1
>> b = myBinAdder ([1 1 0],[1 0 1])
b =
     1     0     1     1
```

5. What is the effect of allocating more bits to the fraction versus the characteristic and vice versa? What is the effect of allocating more bits to the sign?

m **6.** Write a function with header [d] = myIEEE2Dec (IEEE), where IEEE is a 1 × 32 array of ones and zeros representing a 32-bit IEEE754 number. The output should be d, the equivalent decimal representation of IEEE. The input variable IEEE will always be a 32-element array of ones and zeros defining a 32-bit single precision float.

Test Cases:

```
>> IEEE = [1 1 0 0 0 0 1 0 0 1 0 0 0 0 0 0 0 0 0 0 0 0 0 0 0 0 0 0 0 0 0 0];
>> d = myIEEE2Dec (IEEE)
d =
   -48
>>IEEE = [ 1  1  0  0  0  1  0  1  0  0  0  0  0  1  0  1  0  0  0  0  1  0
 0  0  1  0  1  1  1  0  0  0 ];
>> d= myIEEE2Dec (IEEE)
d =
   -2.128544921875000e+003
>> IEEE = zeros (1,32);
>> d = myIEEE2Dec (IEEE)
d =
   5.8775e-039
>> IEEE = ones (1,32);
>> d = myIEEE2Dec (IEEE)
d =
   -6.8056e+038
```

.m 7. Write a function with header `[IEEE] = myDec2IEEE(d)`, where d is a number in decimal and IEEE a 1×32 array of ones and zeros representing the 32-bit IEEE754 closest to d. You can assume that d will not cause an overflow for 32-bit IEEE numbers.
Test Cases:

```
>> IEEE= myDec2IEEE(15.18484199625)
IEEE =
       0 1 0 0 0 0 0 1 0 1 1 1 0 0 1 0 1 1 1 1 0 1 0 1 0 0 0 1 1 1 0 1
>> IEEE = myDec2IEEE(−309.141740)
IEEE =
       1 1 0 0 0 0 1 1 1 0 0 1 1 0 1 0 1 0 0 1 0 0 1 0 0 0 1 0 0 1 0 1
>> IEEE = myDec2IEEE(−25252)
IEEE =
       1 1 0 0 0 1 1 0 1 1 0 0 0 1 0 1 0 1 0 0 1 0 0 0 0 0 0 0 0 0 0 0
```

.m 8. Write a function with header `[] = myFibTimer()`. This function should store the time required to compute the nth Fibonacci number for $n = 5, 6, \cdots, 20$ using the recursive and iterative implementation of Fibonacci numbers (both of which can be found in the reader). The function should produce a plot with the Fibonacci number on the x-axis and the run time on the y-axis. We know that this function will produce different results, even when run multiple times on the same computer. Don't worry. Just get the right idea on the plot. Put a title and axis labels on your plot. The grid and legend are NOT necessary, nor do the tick marks on the x- and y-axis have to be exactly the same.

Hint: The functions `tic` and `toc` will be useful.

Test Cases:

```
>> myFibTimer
```

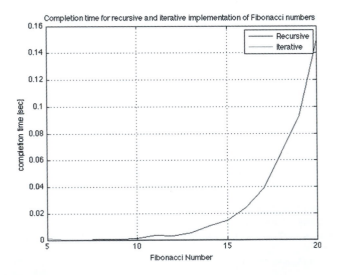

9. Define IEEEBaby to be a representation of numbers using 6 bits where the first bit is the sign bit, the second and third bits are allocated to the characteristic, and the fourth, fifth, and sixth bits are allocated to the fraction. The normalization for the characteristic is 1.

    Write all the decimal numbers that can be represented by IEEEBaby.
    What is the largest/smallest gap in IEEEBaby?

10. Use the `eps` function to determine the smallest number such that the gap is 1.

11. What are some of the advantages and disadvantages of using binary versus decimal?

12. Write the number 13 (base10) in base1. How would you do addition and multiplication in base1?

13. How high can you count on your fingers if you count in binary?

14. Let b be a binary number having n digits. Can you think of ways to multiply and divide b by 2 that does not involve any arithmetic? Hint: Think about how you multiply and divide a decimal number by 10.

Errors, Good Programming Practices, and Debugging

Motivation

Regardless of how proficient, diligent, and careful a programmer you are, writing code with errors is *unavoidable*, and this can be one of the most frustrating parts of programming. As such, dealing with errors preemptively, mentally, and emotionally is a critical part of becoming a proficient programmer. In this chapter, we give a formal definition of errors, provide good programming practices that will help you avoid making errors, and show you some MATLAB tools to help you find errors when you make them.

9.1 Error Types

There are three basic types of errors that programmers need to be concerned about: **syntax errors**, **runtime errors**, and **logical errors**. **Syntax** is the set of rules that govern a language. In written and spoken language, rules can be bent or broken to accommodate the speaker or writer. However, in a programming language the rules are completely rigid. A syntax error occurs when the programmer writes an instruction using incorrect syntax. For example, `1 = x` is not legal in the MATLAB programming language because numbers cannot be assigned as variables. If the programmer tries to execute one of these instructions or any other syntactically incorrect statement, MATLAB will return an error to the programmer in the form of a red message with the line where the error occurred and the probable cause. This is commonly called **throwing** an error.

An Introduction to MATLAB® Programming and Numerical Methods. http://dx.doi.org/10.1016/B978-0-12-420228-3.00009-9

EXAMPLE: Syntax error examples.

```
>> 1 = x
??? 1 = x
     |
Error: The expression to the left of the equals sign is not a valid target for an
assignment.

>> [(1]
??? [(1]
       |
Error: Unbalanced or unexpected parenthesis or bracket.
```

Syntax errors are usually easily detectable, easily found, and easily fixed. Runtime errors are much more difficult to find. Runtime errors are only detectable when a program is run. For example, concatenation is legal in MATLAB syntax, but if you try to concatenate arrays of incorrect dimensions, then MATLAB will not be able to carry out your instruction, and an error will be produced.

EXAMPLE: Runtime error examples.

```
>> x.y = 2;
>> x + 2
??? Undefined function or method 'plus' for input arguments of type 'struct'.

>> [[2 1], [1;2]]
??? Error using ==> horzcat
CAT arguments dimensions are not consistent.
```

Most runtime errors are also easy to find because MATLAB will stop and tell you where the problem is. After programming a function, seasoned programmers will usually run the function several times, allowing the function to throw any errors so that they can fix them.

One of the most difficult kinds of runtime errors to find is called a **logic error**. A logic error does not throw an error, but is an error because the output you get is not the solution you expect. For example, consider the following erroneous implementation of the factorial function.

EXAMPLE: Erroneous factorial function.

```
function [out] = myBadFactorial(n)
% [out] = myBadFactorial(n)
% Erroneous implementation of factorial
% author
% date
```

```
out = 0;
for i = 1:n
    out = out*i;
end

end % end myBadFactorial
```

This function will not produce an error for any input that is valid for a correctly implemented factorial function. However, if you try using `myBadFactorial` at the command window, you will find that the answer is always 0 because out is initialized to 0 instead of 1. Therefore, the line `out = 0` is a logic error. It does not produce an error by MATLAB, but it leads to an incorrect computation of factorial.

Although this kind of error seems unlikely to occur or at least as easy to find as other kinds of errors, when programs become longer and more complicated, they are very easy to generate and notoriously difficult to find. When logic errors occur, you have no choice but to meticulously comb through each line of your code until you find the problem. For these cases, it is important to know exactly how MATLAB will respond to every command you give and not make any assumptions. You can also use MATLAB's debugger, which will be described in the last section of this chapter.

9.2 Avoiding Errors

There are many techniques that can help prevent errors and make it easier for you to find them when they occur. Becoming familiar with the types of mistakes common in programming is a "learning as you go" process; therefore, we could not possibly list them all here. However, we present a few of them in the following section to help get you started building good habits.

9.2.1 Plan your program

When writing an essay, it is important to have a structure and a direction that you intend to follow. To help make your structure more tangible, writing an essay usually starts with an outline containing the main points you wish to address in your paper. This is even more important to do when programming, because computers are more strict than humans when interpreting what you write. Therefore, for complicated programs you should start with an outline of your program that addresses all the tasks you want your program to perform and in the order in which it should perform them.

Many novice programmers, eager to finish their assignments, will attempt to rush to the programming part without properly planning out the tasks that are needed to accomplish the given task. Haphazard planning results in equally haphazard code that is full of errors. Time spent planning out what you are trying to do will be well spent, and you will surely finish faster than had you thrown together a program.

So what does planning a program consist of? Recall in Chapter 3 that a function is defined as a sequence of instructions designed to perform a certain task. A **module** is a function or group of functions that perform a certain task. It is important to design your program in terms of modules, especially for tasks that need to be repeated over and over again. Each module should accomplish a small, well-defined task and know as little information about other functions as possible (i.e., have a very limited set of inputs and outputs).

A good rule of thumb is to plan from the top to bottom and then program from the bottom to the top. That is, decide what the overall program is supposed to do, then determine what the main tasks are to complete the program, and then break the main tasks into components until the module is small enough that you are confident you can write it without errors.

9.2.2 Test everything often

Along the lines of coding in modules, you should test each module for test cases for which you know the answer and enough cases to be confident that the function is working properly (including corner cases). For example, if you are writing a function that tells you whether a number is prime or not, you should test the function for inputs of 0 (corner case), 1 (corner case), 2 (simple yes), 4 (simple no), and 97 (complicated no). If it passes all the test cases, you can move on to other modules, confident that the current module works correctly. This is especially important if subsequent modules depend on or call the module you are working on. If you assume incorrect code is correct because you did not test it, when you get an error in later modules, you will not know whether the error is in the module you are working on or in a previous module, and this will make finding the error more difficult.

You should also test often, even within a single module or function. When you are working on a particular module that has several steps, you should do intermediate tests to make sure it is correct up to the point you have completed. Then if you ever get an error, it will probably be in the part of your code written since the last time you did test it. Even many seasoned programmers are guilty of writing pages and pages of code without testing and then having to spend hours finding a small error somewhere.

9.2.3 Keep your code clean

Just like good craftsmen keep their work area as clean as possible, free of unnecessary clutter, so should you keep your code as clean as possible. There are many things you can do to keep your code clean. First, you should write your code in the fewest instructions possible. For example, » $y = x^2 + 2*x + 1$; is better than $y = x^2$; $y = y + 2*x$; $y = y + 1$. Even if the outcome is the same, every character you type is a chance that you will make a mistake; therefore, reducing how much you write down reduces your risk. Additionally, writing a complete expression will help you and other people understand what you are doing. In the previous example, in the first case it is clear that you are computing the value of a quadratic at x, while in the second case it is not clear.

You can also keep your code clean by using variables rather than values.

EXAMPLE: Poor implementation of adding 10 random numbers.

```
S = 0;
A = rand(1,10);
for i = 1:10
    S = S + A(i);
end
```

EXAMPLE: Good implementation of adding 10 random numbers.

```
N = 10;
S = 0;
A = rand(1,N);
for i = 1:N
     S = S + A(i);
end
```

The second implementation is better for two reasons: first, it is easier for anyone reading your code that N represents the number of random numbers you want to add up, and it appears rationally where it is supposed to in the code (i.e., when creating the list of random numbers and when indexing the list in the for-loop); second, if you ever wanted to change the number of random numbers to add up, you would only have to change it in one place at the beginning. This reduces the chances of making mistakes while writing the code and when changing the value of N.

Again, this is not critical for such a small piece of code. However, it will become very important when your code becomes more complicated and values must be reused many times.

You can also keep your code clean by giving your variables short, descriptive names. For example, N is a sufficient variable for such a simple task as given earlier. The variable name x is probably a good name since x usually holds value of position, rather than a number. Likewise, theNumberOfRandomNumbersToBeAdded is also a poor variable name even though it is descriptive.

Finally, you can keep your code clean by commenting frequently. Although no commenting is certainly bad practice, over-commenting can be just as bad practice. However, different programmers will disagree on exactly how much commenting is appropriate. It will be up to you to decide what level of commenting is appropriate.

9.3 Try/Catch

Often it is important that a function handle certain types of errors gracefully. More specifically, the error must not cause a critical error that makes your program shut down. A **Try-Catch statement** is a code block that allows your program to take alternative actions in case an error occurs.

CONSTRUCTION: Try-Catch Statement

```
try
     code block 1
catch errorInfo
     code block 2
end
```

MATLAB will first attempt to execute the code in the `try` statement (code block 1). If any error occurs, the code in the `catch` statement will be executed (code block 2). Information about the error that activated the `catch` statement is stored in the `struct errorInfo`. After handling the error as you see fit (in code block 2), you can reinitiate the error using the function `rethrow`.

> **WARNING!** Try-catch statements should never be used in place of good programming practice. For example, you should not code sloppily and then encase your program in a try-catch statement until you have taken every measure you can think of to ensure that your function is working properly.

9.4 Type Checking

MATLAB is a **weakly typed** programming language. This means that any variable can take on any data type at any time. For example, you can write » x = 1; immediately followed by » x = 's'. In **strongly typed** programming languages, you must declare what kind of data type your variable is to have before you use it, and the data type usage of your variable cannot change within the scope of a function. Although it seems inconvenient to have to declare the data type of each of your variables, having a strongly typed language helps ensure that you are not abusing the programming language and that your function is being used properly when it is finished.

In the case of MATLAB, there is no way to ensure that the user of your function is inputting variables of the data type you expect. For example, the function `myAdder` in Chapter 3 is designed to add three numbers together. However, the user can input strings, structs, cells, or function handles, each of which will cause different levels of problems. You can have your function **type check** the input variables before continuing and force an error using the `error` function. The error function takes `sprintf` type inputs.

> **TRY IT!** Modify `myAdder` to type check that the input variables are doubles. If any of the input variables are not doubles, the function should return an appropriate error to the user using the error function. Try your function for erroneous input arguments to verify that they are checked.
>
> ```
> function [out] = myAdder(a,b,c)
> % [out] = myAdder(a,b,c)
> % out is the sum of a,b, and c.
> % author
> % date
>
> % type check
> if ~isnumeric(a) || ~isnumeric(b) || ~isnumeric(c)
>     error('Input arguments must have type double.')
> end
>
> % assign output
> out = a + b + c;
>
> end
> ```

```
>> d = myAdder(1,2,'3')
??? Error using ==> myAdder at 9
Input arguments must have type double.
```

9.5 Debugging

Debugging is the process of systematically removing errors, or bugs, from your code. MATLAB has functionalities that can assist you when debugging. MATLAB's debugger opens when you insert a **breakpoint** into your code. A breakpoint is a line in your code at which MATLAB will stop when the function is run. Figure 9.1 shows a breakpoint put at line 8 of myAdder.

To insert a breakpoint, you can click the button [icon] in the upper right or click the small horizontal line to the left of the line of code where you wish MATLAB to stop. This is shown in Figure 9.1.

When you run your code, the editor will open with an arrow next to the line where MATLAB is stopped (Figure 9.2). You can **step** through your code line by line by pushing the [icon]. At each step you can check the values of all the variables in the function's workspace to make sure they all have the expected value.

You can also **step in** to code. That is, when you encounter a line that calls a function, you will enter the workspace of the function called. You can step by clicking the step in button [icon].

```
 1    function [out] = myAdder(a,b,c)
 2    % [out] = myAdder(a,b,c)
 3    % out is the sum of a,b, and c.
 4    % author
 5    % date
 6
 7    % type check
 8    if ~isnumeric(a) || ~isnumeric(b) || ~isnumeric(c)
 9        error('Input arguments must have type double.')
10    end
11
12    % assign output
13    out = a + b + c;
14
15    end
```

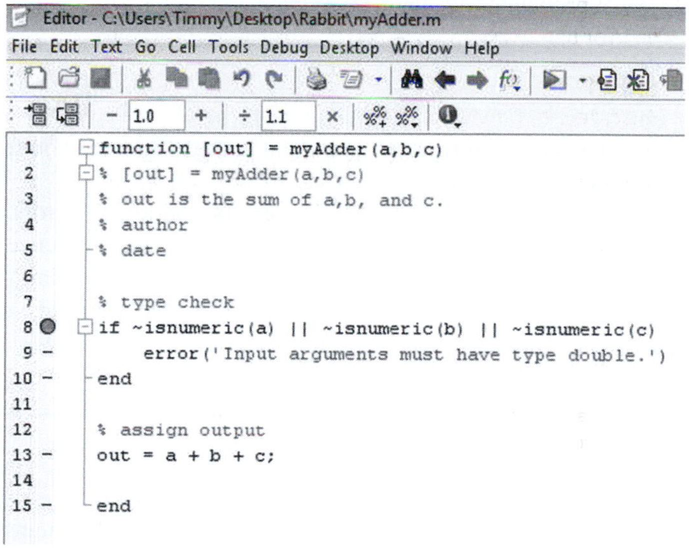

FIGURE 9.1

Breakpoint inserted at line 8.

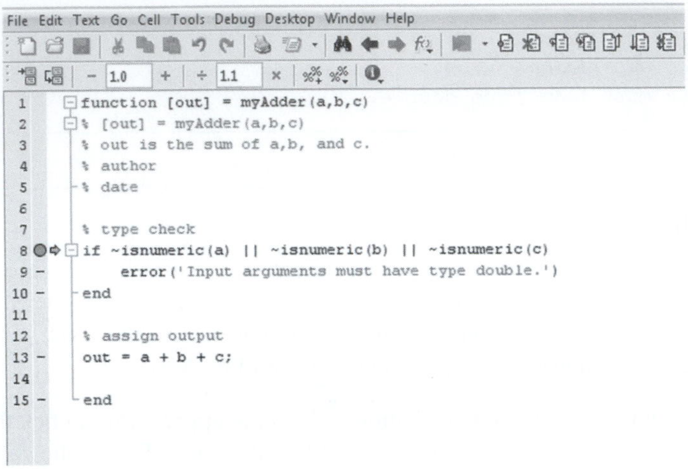

```
File  Edit  Text  Go  Cell  Tools  Debug  Desktop  Window  Help
1    function [out] = myAdder(a,b,c)
2    % [out] = myAdder(a,b,c)
3      % out is the sum of a,b, and c.
4      % author
5      % date
6
7      % type check
8    if ~isnumeric(a) || ~isnumeric(b) || ~isnumeric(c)
9          error('Input arguments must have type double.')
10     end
11
12     % assign output
13     out = a + b + c;
14
15     end
```

FIGURE 9.2

MATLAB stopped at breakpoint at line 8.

When you are finished debugging, you can clear all the breakpoints using the [button] button and exit the debugger by clicking the [button] button.

Using the MATLAB's debugger can be extremely helpful in finding and fixing errors in your code. We encourage you to use the debugger for large programs.

Summary

1. Errors are inevitable when coding. Errors are important because they tell you that something is not working the way you intended.
2. There are three types of errors: syntax errors, runtime errors, and logical errors.
3. You can reduce the numbers of errors in your coding with good coding practice.
4. Try-catch statements can be used to handle unexpected errors without stopping your code. However, try-catch statements should never be used in place of good practice to manage errors.
5. The Debugger is a MATLAB tool for helping you find errors.

Vocabulary

breakpoint	step	throwing an error
debugging	step in	try-catch statement
logic error	strongly typed	type check
module	syntax	weakly typed
runtime error	syntax error	

Functions and Operators

breakpoint	error	rethrow
breakpoint	insert	try
catch	remove	warning

Problems

(none)

Reading and Writing Data

Motivation

Storing data and the results of your programming efforts is important for working over multiple sessions and sharing your results with collaborators. Since when MATLAB closes, all the variables in the workspace are lost, data must be stored in some other way than workspace variables. Sometimes data must also be readable by or written in a form that can be read by other programs.

This chapter shows you how you can use MATLAB to read and write data in several common formats.

10.1 .mat Files

A **.mat file** is the standard format for saving data within MATLAB. These files can be created using the `save` function and loaded using the `load` function. Note that the `save` function will permanently store a .mat file on your hard drive in the current working directory.

> **TRY IT!** Create variables `x = 1`, `y = 'string'`, and `z = 1, 'string'` and save them in the file `test.mat` using the `save` function. Clear the variables from the workspace using `clear`. Recall the variables in `test.mat` using the `load` function and verify that they are in the workspace.

An Introduction to MATLAB® Programming and Numerical Methods. http://dx.doi.org/10.1016/B978-0-12-420228-3.00010-5

```
>> x = 1;
>> y = 'string';
>> z = {1, 'string'};
>> save test x y z
>> clear all
>> load test.mat
>> x
x =
      1
>> y =
string
>> z
z =
    [1]     'string'
```

Note that you do not need to list all the variables in the workspace when using the `save` function. You can just list the ones you want saved for next session. If in the previous example, the line >> `save test` was used in place of >> `save text x y z`, then MATLAB would store all the variables in the workspace in `test.mat`.

10.2 .txt Files

A **.txt** file, or **text file**, is a file containing only plain text. Text files are used to store data, and most programs that work with data can read and write them. However, programs you write and programs that read your text file will usually expect the text file to be in a certain format; that is, organized in a specific way.

To work with text files you must first create a **file identifier** associated with the relevant text file. A file identifier is an integer that MATLAB associates with a text file, similar to the way a function handle is associated with a function. You can create a file identifier using the `fopen` function. The `fopen` function has header `[fid] = fopen(filename, permissions)` where `filename` is a string with the desired name of text file, and `permissions` is a string describing how the relevant text file is going to be used. A list of permissions is given in Table 10.1. You can find the same table in the help for `fopen`.

If you try to use `fopen` to open a file that does not exist or use a permission that does not create the file, you will get a file identifier of -1. Once you have opened a text file for writing, you can write to the file using several functions. For this course, we will use the `fprintf` function. The `fprintf` function works similarly to `sprintf` except that it writes to a file rather than to a string output. You can close the file from writing using `fclose`. After `fclose` is called on a file identifier, it cannot be written to again without discarding the data. The output argument to `fprintf` is the number of bytes written to the text file. You usually do not need to assign this output (i.e., just suppress the line of code).

Table 10.1   Permissions for `fopen` Function

Permission	Description
'r'	Open file for reading
'w'	Open file for writing; discard existing contents
'a'	Open or create file for writing; append data to end of file
'r+'	Open (do not create) file for reading and writing
'w+'	Open or create file for reading and writing; discard existing contents
'a+'	Open or create file for reading and writing; append data to end of file
'W'	Open file for writing without automatic flushing
'A'	Open file for appending without automatic flushing

TRY IT! Create a text file called `test.txt` and populate it with mock values using `fprintf`. Close the file for writing using `fclose`.

```
>> fid = fopen('test.txt', 'w+')
fid =
     3
>> for i = 1:5
fprintf(fid, '%d %d %d\n', i, i^2, i^3);
end
>> fclose(fid);
```

This produces the text file `test.txt` shown in Figure 10.1.

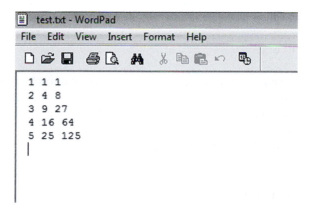

FIGURE 10.1

File `test.txt` opened in wordpad for PC.

You will notice that the file `test.txt` is permanently stored on your computer in the working directory of MATLAB.

There are also several functions for reading data from text files. The easiest to use is probably `fgetl`. To read data from text files you will need to open the file again using `fopen`. The file identifier outputted from `fopen` will be used as input to `fgetl`. Each call to `fgetl` will return the next line from the text file until there are no more lines. Unfortunately each line is returned as a `string`, so if you are reading numeric data, you will have to convert the string to doubles using the `str2num` function. If you have read every line of data contained in the text file, `fgetl` will return the `double` −1.

TRY IT! Read the data from `test.txt` using `fgetl`. Store the values in a 5 × 3 matrix as doubles. Hint: Use a `while` loop.

```
>> fid = fopen('test.txt','r');
>> tline = fgetl(fid);
>> M = [];
>> while ~isnumeric(tline)
M = [M; str2num(tline)];
tline = fgetl(fid);
end
>> M
M =
        1       1       1
        2       4       8
        3       9      27
        4      16      64
        5      25     125
```

10.3 .xls Files

You can also read and write out of Microsoft Excel files, which is a useful format for dealing with large tables of data. The remainder of this section assumes some working knowledge of Microsoft Excel. If you do not know about Microsoft Excel, it is widely available, and should be both searchable on the Internet and installed on most campus computers in the United States.

Writing to an Excel file is done using the function `xlswrite`. The `xlswrite` function has header `[success,message]=xlswrite(file,array,sheet,range)`, although it is not always necessary to use all the inputs and outputs. The input variable `file` is the name of the `.xls` file that is to be written to. If no such file exists, then the file will be created in the working directory. The input variable `array` is a double array or a cell array. Each element of `array` will be written to a single cell in an Excel spreadsheet (assuming that the elements of the cell are doubles or a `chars`).

The last two input variables are optional (i.e., they can be omitted). The input variable `sheet` is the name of the sheet the data is to be written to. If the sheet does not exist, then it will be created in the `.xls` file. The last input variable `range` is a string describing the location in the sheet where the data is to be written. Microsoft Excel has an alphanumeric numbering system, where rows are denoted by numbers and columns are denoted by letters. Column values are given first. The range is specified by the upper left coordinate and the lower right coordinate, separated by a comma. So A1:J10 would go from the first row, first column to the tenth row, tenth column.

The output argument of `xlswrite`, `success`, is 1 if the data in the array was written successfully, and 0 otherwise. The output variable, `message`, is a struct containing information about what happened. You usually do not need to assign this output.

TRY IT! Use `xlswrite` to populate an Excel file called `test.xls` with mock values (see Figure 10.2).

```
>> A = rand(100,10);
>> [S] = xlswrite('test.xls', A)
S =
     1
```

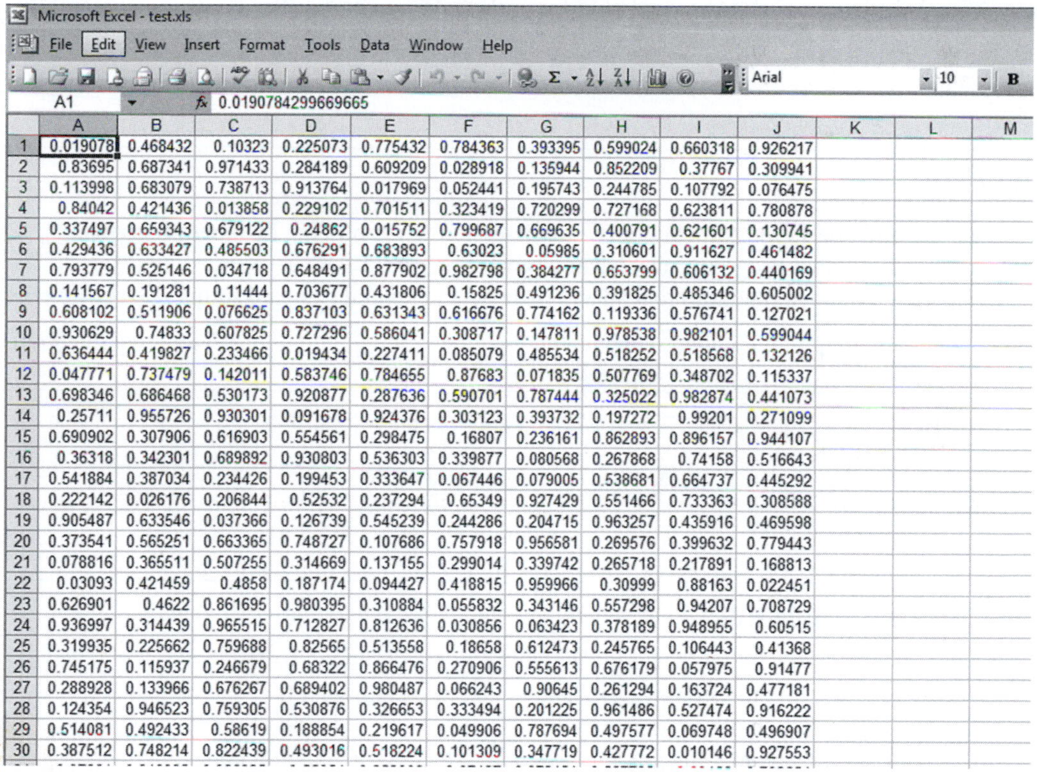

	A	B	C	D	E	F	G	H	I	J	K	L	M
1	0.019078	0.468432	0.10323	0.225073	0.775432	0.784363	0.393395	0.599024	0.660318	0.926217			
2	0.83695	0.687341	0.971433	0.284189	0.609209	0.028918	0.135944	0.852209	0.37767	0.309941			
3	0.113998	0.683079	0.738713	0.913764	0.017969	0.052441	0.195743	0.244785	0.107792	0.076475			
4	0.84042	0.421436	0.013858	0.229102	0.701511	0.323419	0.720299	0.727168	0.623811	0.780878			
5	0.337497	0.659343	0.679122	0.24862	0.015752	0.799687	0.669635	0.400791	0.621601	0.130745			
6	0.429436	0.633427	0.485503	0.676291	0.683893	0.63023	0.05985	0.310601	0.911627	0.461482			
7	0.793779	0.525146	0.034718	0.648491	0.877902	0.982798	0.384277	0.653799	0.606132	0.440169			
8	0.141567	0.191281	0.11444	0.703677	0.431806	0.15825	0.491236	0.391825	0.485346	0.605002			
9	0.608102	0.511906	0.076625	0.837103	0.631343	0.616676	0.774162	0.119336	0.576741	0.127021			
10	0.930629	0.74833	0.607825	0.727296	0.586041	0.308717	0.147811	0.978538	0.982101	0.599044			
11	0.636444	0.419827	0.233466	0.019434	0.227411	0.085079	0.485534	0.518252	0.518568	0.132126			
12	0.047771	0.737479	0.142011	0.583746	0.784655	0.87683	0.071835	0.507769	0.348702	0.115337			
13	0.698346	0.686468	0.530173	0.920877	0.287636	0.590701	0.787444	0.325022	0.982874	0.441073			
14	0.25711	0.955726	0.930301	0.091678	0.924376	0.303123	0.393732	0.197272	0.99201	0.271099			
15	0.690902	0.307906	0.616903	0.554561	0.298475	0.16807	0.236161	0.862893	0.896157	0.944107			
16	0.36318	0.342301	0.689892	0.930803	0.536303	0.339877	0.080568	0.267868	0.74158	0.516643			
17	0.541884	0.387034	0.234426	0.199453	0.333647	0.067446	0.079005	0.538681	0.664737	0.445292			
18	0.222142	0.026176	0.206844	0.52532	0.237294	0.65349	0.927429	0.551466	0.733363	0.308588			
19	0.905487	0.633546	0.037366	0.126739	0.545239	0.244286	0.204715	0.963257	0.435916	0.469598			
20	0.373541	0.565251	0.663365	0.748727	0.107686	0.757918	0.956581	0.269576	0.399632	0.779443			
21	0.078816	0.365511	0.507255	0.314669	0.137155	0.299014	0.339742	0.265718	0.217891	0.168813			
22	0.03093	0.421459	0.4858	0.187174	0.094427	0.418815	0.959966	0.30999	0.88163	0.022451			
23	0.626901	0.4622	0.861695	0.980395	0.310884	0.055832	0.343146	0.557298	0.94207	0.708729			
24	0.936997	0.314439	0.965515	0.712827	0.812636	0.030856	0.063423	0.378189	0.948955	0.60515			
25	0.319935	0.225662	0.759688	0.82565	0.513558	0.18658	0.612473	0.245765	0.106443	0.41368			
26	0.745175	0.115937	0.246679	0.68322	0.866476	0.270906	0.555613	0.676179	0.057975	0.91477			
27	0.288928	0.133966	0.676267	0.689402	0.980487	0.066243	0.90645	0.261294	0.163724	0.477181			
28	0.124354	0.946523	0.759305	0.530876	0.326653	0.333494	0.201225	0.961486	0.527474	0.916222			
29	0.514081	0.492433	0.58619	0.188854	0.219617	0.049906	0.787694	0.497577	0.069748	0.496907			
30	0.387512	0.748214	0.822439	0.493016	0.518224	0.101309	0.347719	0.427772	0.010146	0.927553			

FIGURE 10.2

File from previous example, `test.xls`, opened in Microsoft Excel.

You can read data contained in `.xls` files using the `xlsread` function. The `xlsread` function has header `[numeric,txt,raw]=xlsread(file,sheet,range)`. The string input arguments `file`, `sheet`, and `range` are the filename, file, the sheet name, sheet, and the position of the cells desired using Microsoft Excel notation, respectively. The numeric data is contained in the output argument, `numeric (double)`, the string data is contained in the cell array `txt`, and all the data that MATLAB could not process is contained in the cell array `raw`. If you know beforehand that there is no other data besides numeric data, you can use only the output argument `numeric`.

TRY IT! Use the `xlsread` function to read the upperleft most 5×5 matrix of data values contained in `test.xls`. Store the retrieved data in the matrix, A.

```
>> A = xlsread('test.xls', 'Sheet1', 'A1:E5')
A =
      0.0191    0.4684    0.1032    0.2251    0.7754
      0.8370    0.6873    0.9714    0.2842    0.6092
      0.1140    0.6831    0.7387    0.9138    0.0180
      0.8404    0.4214    0.0139    0.2291    0.7015
      0.3375    0.6593    0.6791    0.2486    0.0158
```

Summary

1. Data must often be stored for a later MATLAB session or for reading by other programs.
2. Data created by other programs may have to be read by MATLAB.
3. MATLAB has built-in functions to read and write data in several standard forms: `.mat`, `.txt`, and `.xls`.

Vocabulary

file identifier	text file	.txt file
.mat file		.xls file

Functions and Operators

fclose	fread	xlsread
fgetl	fscanf	xlswrite
fopen	load	
fprintf	save	

Problems

(none)

Visualization and Plotting

CHAPTER OUTLINE

Visualizing data is usually the best way to convey important engineering ideas and information, especially if the information is made up of many, many numbers. The ability to visualize and plot data quickly and in many different ways is one of MATLAB's most powerful features.

MATLAB has numerous graphics generators that enable you to efficiently display plots, surfaces, volumes, vector fields, histograms, animations, and many other data plots. By the end of this chapter, you should be familiar with the most common ones and have enough information to explore the rest.

It is worth noting that MATLAB's graphical interface utilizes a style of programming called Object Oriented Programming (OOP), which is a different school of thought than the type of programming presented in this book. Thus we have tried to show you everything you need to know to use these features without going into detail about how they work.

11.1 2D Plotting

The basic plotting function in MATLAB is plot (x,y). The plot function takes in two arrays, x and y, and produces a visual display of the respective points in x and y.

TRY IT! Given the arrays x = [0 1 2 3] and y = [0 1 4 9], use the plot function to produce a plot of x versus y.

```
>> x = [0 1 2 3];
>> y = [0 1 4 9];
>> plot(x,y)
```

You will notice in Figure 11.1 that by default, the plot function connects each point with a blue line. To make the function look smooth, use a finer discretization of points.

An Introduction to MATLAB® Programming and Numerical Methods. http://dx.doi.org/10.1016/B978-0-12-420228-3.00011-7

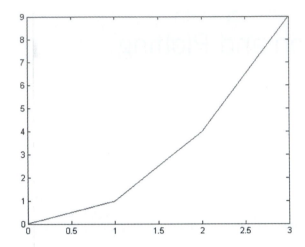

FIGURE 11.1

Example of plot (x,y) where x and y are vectors.

TRY IT! Make a plot of the function $f(x) = x^2$ for $-5 \le x \le 5$ (Figure 11.2).

```
>> x = linspace(-5,5,100);
>> plot(x,x.^2)
```

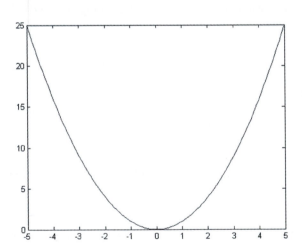

FIGURE 11.2

Example of plot of the function $f(x) = x^2$ on the interval $[-5,5]$.

Table 11.1 Color and Line Style Symbols

Symbol	Description
b	blue
g	green
r	red
c	cyan
m	magenta
y	yellow
k	black
w	white
.	point
o	circle
x	x-mark
+	plus
*	star
s	square
d	diamond
v	triangle (down)
^	triangle (up)
<	triangle (left)
>	triangle (right)
p	pentagram
h	hexagram
–	solid
:	dotted
–.	dashdot
– –	dashed
(none)	no line

To change the marker or line, you can put a third input argument into `plot`, which is a string that denotes the color and line style to be used in the plot. For example, `plot (x,y,'ro')` will plot the elements of x against the elements of y using red, r, circles, `'o'`. The possible marker colors and sizes are shown in Table 11.1.

TRY IT! Make a plot of the function $f(x) = x^2$ for $-5 \leq x \leq 5$ using a dashed green line (Figure 11.3).

```
>> x = linspace(−5,5,100);
>> plot(x,x.^2,'g− −')
```

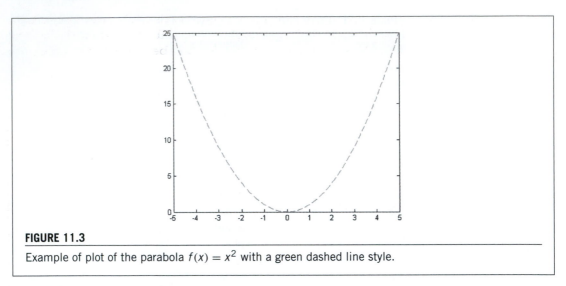

FIGURE 11.3

Example of plot of the parabola $f(x) = x^2$ with a green dashed line style.

You can plot more than one data set into a single graph using the hold on command. Typing hold on will put all subsequent plots on the same graph until hold off is typed.

TRY IT! Make a plot of the function $f(x) = x^2$ and $g(x) = x^3$ for $-5 \le x \le 5$ (Figure 11.4). Use different colors and markers for each function.

```
>> x = linspace(-5,5,20);
>> hold on
>> plot(x,x.^2,'ko')
>> plot(x,x.^3, 'r*')
>> hold off
```

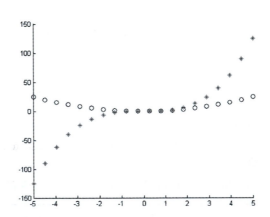

FIGURE 11.4

Other examples of plotting styles, illustrated for $f(x) = x^2$ and $f(x) = x^3$, respectively.

It is customary in engineering to *always* give your plot a title and axis labels so that people know what your plot is about. You can add a title to your plot using the `title` function, which takes as input a `string` and puts that string as the title of the plot. The functions `xlabel` and `ylabel` work in the same way to name your axis labels.

TRY IT! Add a title and axis labels to the previous plot (Figure 11.5).

```
>> x = linspace(-5,5,20);
>> hold on
>> plot(x,x.^2,'ko')
>> plot(x,x.^3, 'r*')
>> hold off
>> title('Plot of Various Polynomials')
>> xlabel('x axis')
>> ylabel('y axis')
```

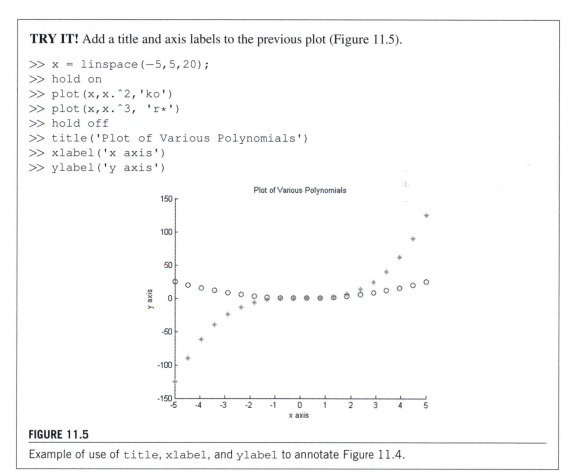

FIGURE 11.5

Example of use of `title`, `xlabel`, and `ylabel` to annotate Figure 11.4.

TIP! You can use `sprintf` function with the title function to make customized titles. For example, you may want to include data-specific content in your title.

TRY IT! Use the `sprintf` function to change the title of the previous plot to "Plot of Various Polynomials from −5 to 5" (Figure 11.6).

```
>> x = linspace(-5,5,20);
>> hold on
>> plot(x,x.^2,'ko')
>> plot(x,x.^3, 'r*')
>> hold off
>> title(sprintf('Plot of Various Polynomials from %d to %d', x(1), x(end)))
>> xlabel('x axis')
>> ylabel('y axis')
```

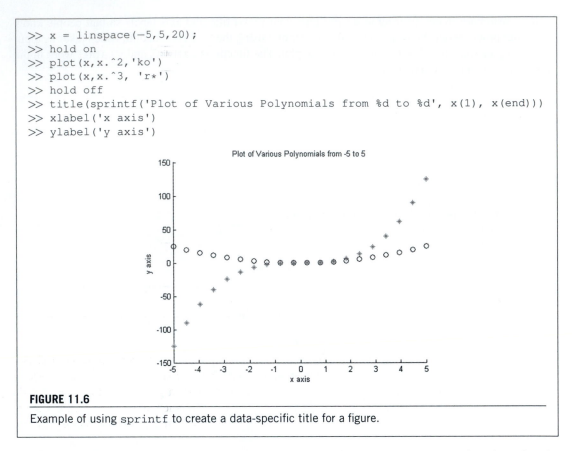

FIGURE 11.6

Example of using `sprintf` to create a data-specific title for a figure.

You can add a legend to your plot by using the `legend` function. The `legend` function takes the same number of strings as input as the number of data sets that are being plotted. It is up to you to put the same number of `legend` entries as data sets. Too few will leave some of your data sets unlabeled, and too many will ignore the extras and give you a warning.

TRY IT! Add a legend to the previous plot using the `legend` function (Figure 11.7).

```
>> x = linspace(-5,5,20);
>> hold on
>> plot(x,x.^2,'ko')
>> plot(x,x.^3, 'r*')
>> hold off
>> title(sprintf('Plot of Various Polynomials from %d to %d', x(1), x(end)))
>> xlabel('x axis')
>> ylabel('y axis')
>> legend('quadratic','cubic')
```

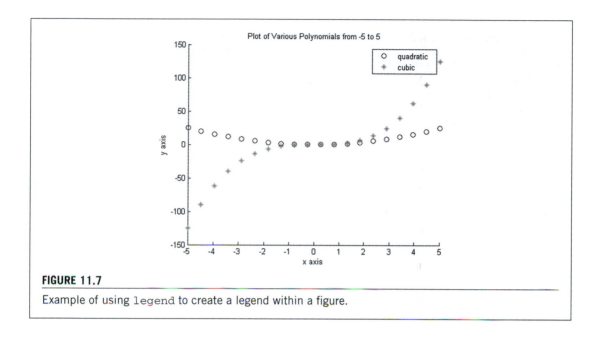

FIGURE 11.7

Example of using `legend` to create a legend within a figure.

Finally, you can further customize the appearance of your plot using the `axis` function and the `grid` command. The `axis` function takes in a 1 × 4 array of the form [xmin xmax ymin ymax], which denotes the limits of each axis. The `grid on` command adds a grid to the axis, and `grid off` removes it. You can also use options with the `axis` command, such as `axis equal`, `axis square`, `axis tight`, and several others. The description for these options can be found in the help function for the `axis` function.

TRY IT! Use the `axis` function to change the limits such that x is visible from −6 to 6 and y is visible from −10 to 10 (Figure 11.8). Turn the grid on.

```
>> x = linspace(-5,5,100);
>> hold on
>> plot(x,x.^2,'ko')
>> plot(x,x.^3, 'r*')
>> hold off
>> title(sprintf('Plot of Various Polynomials from %d to %d', x(1), x(end)))
>> xlabel('x axis')
>> ylabel('y axis')
>> legend('quadratic','cubic')
>> axis([-6 6 -10 10])
>> grid on
```

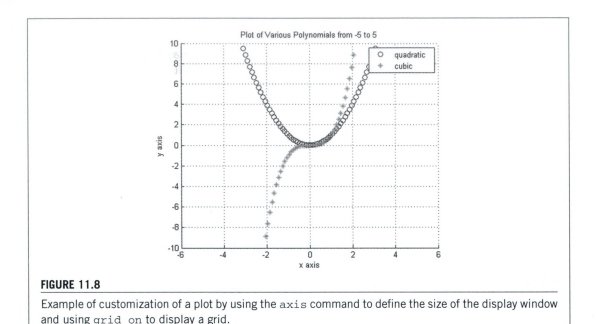

FIGURE 11.8

Example of customization of a plot by using the `axis` command to define the size of the display window and using `grid on` to display a grid.

To open a new figure without overwriting the current figure, type » `figure`. To the close all the current figures, type » `close all`. To clear the contents of a figure without closing the figure, type » `clf`.

You can create a table of plots on a single figure using the `subplot` function. The `subplot` function takes three inputs: the number of rows of plots, the number of columns of plots, and to which plot all calls to plotting functions should plot. You can move to a different `subplot` by calling the `subplot` again with a different entry for the plot location.

There are several other plotting functions that plot x versus y data. Some of them are `scatter`, `bar`, `loglog`, `semilogx`, and `semilogy`. `scatter` works exactly the same as `plot` except it defaults to red circles (i.e., `plot (x,y,'ro')` is equivalent to `scatter (x,y)`). The `bar` function plots bars centered at x with height y. The `loglog`, `semilogx`, and `semilogy` functions plot the data in x and y with the x and y axis on a log scale, the x axis on a log scale and the y axis on a linear scale, and the y axis on a log scale and the x axis on a linear scale, respectively.

TIP! When making complicated plots, it is useful to make a script file that generates the plot rather than generating it from the command prompt. If and when you make a mistake, you will have to reenter many commands, each of which will give you new opportunities to make other mistakes. With a script file, you will be able to change the erroneous line and then rerun the script to make the plot.

TRY IT! Given the arrays x = [0 1 2 3 4 5] and y = [1 2 4 8 16 2], create a 2×3 subplot where each subplot plots x versus y using `plot`, `scatter`, `bar`, `loglog`, `semilogx`, and `semilogy`. Title and label each plot appropriately. Use a grid, but a legend is not necessary. (See Figure 11.9.)

```
%% Script for generating subplots of various
% author
% date

%% clean start
clc
clear
close all

%% define data
x = [0 1 2 3 4 5];
y = [1 2 4 8 16 2];

%% create subplot
subplot(2,3,1)
plot(x,y)
title('Plot')
xlabel('x')
ylabel('y')
grid on

subplot(2,3,2)
scatter(x,y)
title('Scatter')
xlabel('x')
ylabel('y')
grid on

subplot(2,3,3)
bar(x,y)
title('Bar')
xlabel('x')
ylabel('y')
grid on

subplot(2,3,4)
loglog(x,y)
title('Loglog')
xlabel('x')
ylabel('y')
grid on
```

```
subplot(2,3,5)
semilogx(x,y)
title('Semilogx')
xlabel('x')
ylabel('y')
grid on

subplot(2,3,6)
semilogy(x,y)
title('Semilogy')
xlabel('x')
ylabel('y')
grid on
```

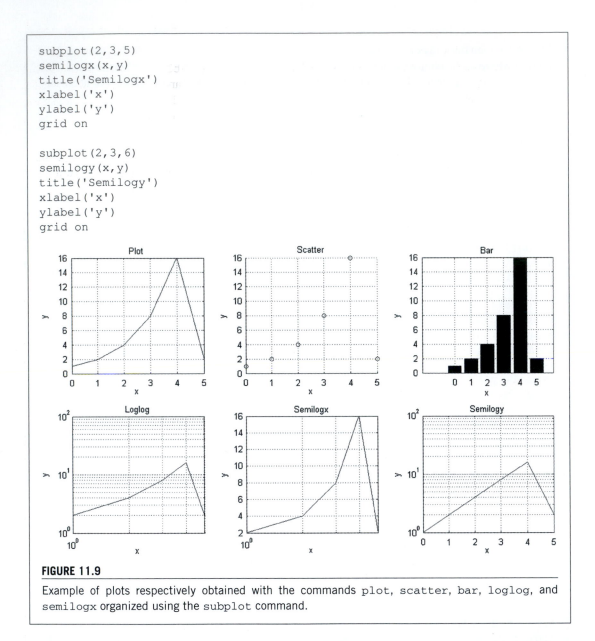

FIGURE 11.9

Example of plots respectively obtained with the commands plot, scatter, bar, loglog, and semilogx organized using the subplot command.

Finally, there are other functions for plotting data in 2D. The errorbar function plots x versus y data but with error bars for each element. The polar function plots θ versus r rather than x versus y. The stemplot function plots stems at x with height at y. The hist function makes a histogram of a data set; X; boxplot gives a statistical summary of a data set; and pie makes a pie chart. The usage of these functions are left to your exploration.

11.2 **3D Plotting**

The basic 3D plotting function is plot3, which has header [] = plot3(x,y,z) where x, y, and z are vectors. The plot3 function will plot all (x,y,z) coordinates, and the default call to plot3 connects subsequent points with a blue line. Commands such as grid, hold, axis, title, xlabel, ylabel, legend, and subplot work the same as when plotting in two dimensions.

TRY IT! Consider the parameterized data set t = [0:pi/50:10*pi], x = sin (t), and y = cos (t). Make a three-dimensional plot of the (x, y, t) data set using plot3. Turn the grid on, make the axis equal, and put axis labels and a title. (See Figure 11.10.)

```
t = [0:pi/50:10*pi];
plot3(sin(t),cos(t),t)
grid on
axis square
title('Parametric Curve')
xlabel('x')
ylabel('y')
zlabel('t')
```

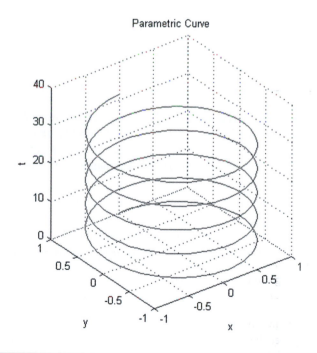

FIGURE 11.10

Example of a three-dimensional plot obtained for the helix ($sin(t)$, $cos(t)$, t) using plot3.

Many times we would like a surface plot rather than a line plot when plotting in three dimensions. In three-dimensional surface plotting, we wish to make a graph of some relationship $f(x, y)$. In surface plotting all (x,y) pairs must be given. This is not straightforward to do using vectors. Therefore, in surface plotting, the first data structure you must create is called a **mesh**. Given vectors of x and y values, a mesh is a listing of all the possible combinations of x and y. In MATLAB, the mesh is given as two matrices X and Y where X (i,j) and Y (i,j) define possible (x,y) pairs. A third matrix, Z, can then be created such that Z (i,j) = f $(X (i,j)$, Y $(i,j))$. A mesh can be created using the meshgrid function in MATLAB. The meshgrid function has header $[X,Y]$ = meshgrid (x,y), where x and y are vectors containing the independent data set. The output variables X and Y are as described earlier.

TRY IT! Create a mesh of the vectors x = [1 2 3 4] and y = [3 4 5] using the meshgrid function.

```
>> x = [1 2 3 4];
>> y = [3 4 5];
>> [X,Y] = meshgrid(x,y)
X =
     1     2     3     4
     1     2     3     4
     1     2     3     4
Y =
     3     3     3     3
     4     4     4     4
     5     5     5     5
```

There are several functions in MATLAB to plot surfaces. Each function has a different look, but they have the same basic function header [] = surf (X,Y,Z), where X and Y are the output arrays from meshgrid, and Z = f (X,Y) or Z (i,j) = f $(X (i,j),Y (i,j))$. The most common surface plotting functions are surf and contour. All commands such as hold, axis, grid, among others work with these plots as well.

TRY IT! Make a 1 × 2 subplot of the surface $f(x, y) = \sin(x) \cdot \cos(y)$ for $-5 \le x \le 5, -5 \le y \le 5$, the first using the surf function and the second using the contour function. Take care to use a sufficiently fine discretization in x and y to make the plot look smooth. (See Figure 11.11.)

```
>> x = -5:.2:5;
>> y = -5:.2:5;
>> [X,Y] = meshgrid(x,y);
>> Z = sin(X).*cos(Y);
>> subplot(1,2,1)
>> surf(X,Y,Z)
>> title('Surface Plot using surf')
>> xlabel('x')
```

```
>> ylabel('y')
>> zlabel('z')
>> subplot(1,2,2)
>> contour(X,Y,Z)
>> title('Surface Plot using contour')
>> xlabel('x')
>> ylabel('y')
```

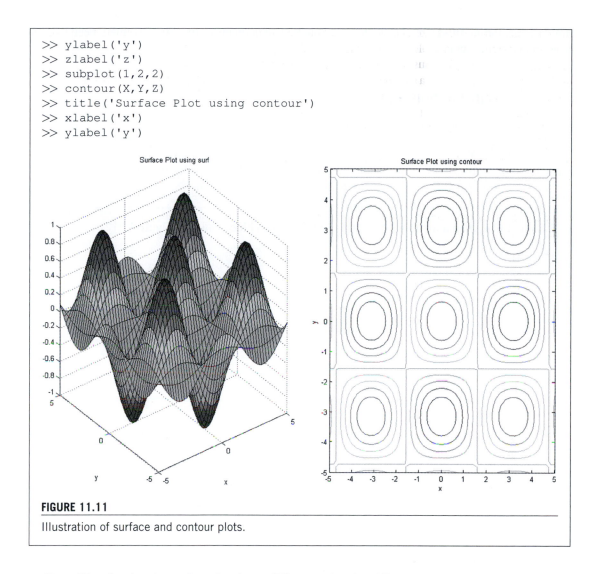

FIGURE 11.11

Illustration of surface and contour plots.

You will notice that the surface plot shows different colors for different elevations, red for higher and blue for lower. The color scheme for a surface plot can be changed using the functions such as `caxis` and `colormap`. These are left as exercises.

There are many more functions related to plotting in MATLAB and this is in no way an exhaustive list. However, it should be enough to get you started so that you can find the plotting functions in MATLAB that suit you best and provide you with enough background to learn how to use them when you encounter them. For example, `bar3` and `isosurface` are more advanced 3D plotting functions that are given as problems. Also, MATLAB's visualization toolbox is very advanced, and explaining all the plot customization methods is beyond the scope of this book. However, reading MATLAB's help for various plotting functions is enough to find out how to do many of the things you will need to do in your engineering careers.

11.3 Animations and Movies

An **animation** is a sequence of still frames, or plots, that are displayed in fast enough succession to create the illusion of continuous motion. Animations and movies often convey information better than individual plots. You can create animations in MATLAB by calling a plot function inside of a loop (usually a for-loop). However, due to the way in which MATLAB handles plotting, you must call the drawnow command for the updated plot to be displayed to the screen.

TRY IT! Create an animation of a red circle following a blue sin wave (Figure 11.12).

```
%% Sin Wave Animation
clc
clear
close all

%% assign parameters
N = 1000;

%% assign x and y coordinates
x = linspace(0,6*pi,N);
y = sin(x);

for i = 1:N

    % create figure
    clf
    hold on
    plot(x,y)
    plot(x(i),y(i),'ro')
    grid on
    axis tight

    title(sprintf('Sin Wave at (%f, %f)', x(i), y(i)))
    xlabel('x')
    ylabel('sin(x)')

    drawnow

end % end for i
```

You can store movies using MATLAB's avifile function. The avifile function has header aviobj = avifile (filename). The input variable filename is a string containing the desired name of the movie file. The output variable aviobj is a variable with a new kind of data type called avifile. A variable with data type avifile contains data about the .avi file being

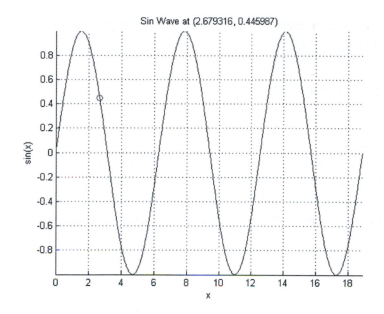

FIGURE 11.12

Snapshot from the animation obtained by execution of the code above.

created. You can store a figure as a frame (another data type) using the `getframe` function and add individual frames to the `avifile` variable using the function `addframe`. Note that the function `gcf` is a reference to the current figure. A more detailed explanation of what this means is beyond the scope of this book. However, you can study this in more detail in books on object oriented programming. When you are done adding frames to the `avifile`, you must close the `avifile` variable using the `close` function. You will see a movie file permanently stored in the working directory of MATLAB after you are finished.

TRY IT! Add code to the previous example so that the movie is stored as an `avi` file called `'test.avi'`.

```
%% Sin Wave Animation
clc
clear
close all

%% assign parameters
N = 1000;

%% assign x and y coordinates
x = linspace(0,6*pi,N);
y = sin(x);
```

```
%% create avi file variable
aviobj = avifile('test');

for i = 1:N
    % create figure
    clf
    hold on
    plot(x,y)
    plot(x(i),y(i),'ro')
    grid on
    axis tight

    title(sprintf('Sin Wave at (%f, %f)', x(i), y(i)))
    xlabel('x')
    ylabel('sin(x)')

    drawnow

    % get figure as frame
    F = getframe(gcf);

    % add frame to avi object.
    aviobj = addframe(aviobj, F);

end % end for i

close(aviobj)
```

Note that this code was run on a PC (i.e., Windows) and may have problems working on a Mac.

Summary

1. Visualizing data is an essential tool in engineering.
2. MATLAB has a vast library of plotting tools that can be used to visualize data.

Vocabulary

animation
mesh

Functions and Operators

addframe	grid minor	semilogx
avifile	hist	semilogy
axis	hold off	stem
bar	hold on	subplot
bar3	legend	surf
box	loglog	surface
close all	mesh	title
contour	meshgrid	view
contourf	patch	waterfall
errorbar	pie	xlabel
figure	plot	ylabel
getframe	plot3	
grid	polar	

Problems

⊠ **1.** A cycloid is the curve traced by a point located on the edge of a wheel rolling along a flat surface. The (x, y) coordinates of a cycloid generated from a wheel with radius, r, can be described by the parametric equations:

$$x = r(\phi - \sin \phi)$$

$$y = r(1 - \cos \phi)$$

where ϕ is the number of radians that the wheel has rolled through.
Generate a plot of the cycloid for $0 \le \phi \le 2\pi$ using 1000 increments and $r = 3$. Give your plot a title and labels. Turn the grid on and modify the axis limits to make the plot neat.

⊠ **2.** Consider the following function:

$$y(x) = \sqrt{\frac{100(1 - 0.01x^2)^2 + 0.02x^2}{(1 - x^2)^2 + 0.1x^2}}.$$

Generate a 2×2 subplot of $y(x)$ for $0 \le x \le 100$ using plot, semilogx, semilogy, and loglog. Use a fine enough discretization in x to make the plot appear smooth. Give each plot axis labels and a title. Turn the grid on. Which plot seems to convey the most information?.

⊠ **3.** Plot the functions $y_1(x) = 3 + \exp{-x} \sin (6x)$ and $y_2(x) = 4 + \exp(-x) \cos (6x)$ for $0 \le x \le 5$ on a single axis using the hold command. Give the plot axis labels, a title, and a legend.

⊠ **4.** Generate 1000 normally distributed random numbers using the randn function. Look up the help for the hist function. Use the hist function to plot a histogram of the randomly generated numbers. Use the hist function with header [N,X] = hist (Y) to distribute the randomly generated numbers into 10 bins. Create a bar graph of output of hist using the bar function. It should look very similar to the plot produced by hist.

Do you think that the randn function is a good approximation of a normally distributed number?.

▷▷ **5.** Let the number of students with A's, B's, C's, D's, and F's be contained in the array gradeDist = [42 85 67 20 5]. Use the pie function to generate a pie chart of gradeDist. Put a title and legend on the pie chart.

▷▷ **6.** Let $-4 \leq x \leq 4$, $-3 \leq y \leq 3$, and $z(x, y) = \frac{xy(x^2-y^2)}{x^2+y^2}$. Create vectors x and y with 100 evenly spaced points over the interval. Create meshgrids X and Y for x and y using the meshgrid function. Compute the matrix Z from X and Y. Create a 2 × 2 subplot where the first row is the surface Z plotted using surf and a plot using mesh, respectively. The second row of the subplot should be the surface Z plotted using the functions contour, and contourf, respectively. Give each axis a title and axis labels.

.m **7.** Write a function with header [] = myPolygon (n) that plots a regular polygon with n sides and radius 1. Recall that the radius of a regular polygon is the distance from its centroid to the vertices. Use axis equal to make the polygon look regular. Remember to give the axes a label and a title. You can use title and sprintf to title the plot according to the number of sides. Use the axis function to make the x-axis and y-axis go from −1 to 1. Hint: This problem is significantly easier if you think in polar coordinates. Recall that a complete revolution around the unit circle is 2π radians. Note: The first and last point on the polygon should be the point associated with the polar coordinate angles, 0 and 2π, respectively. (See Figure 11.13.)

Test Cases:
```
>> myPolygon (5)
```

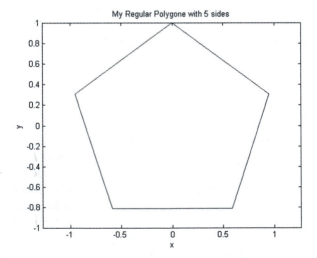

FIGURE 11.13

Test case for problem.m₇ (plotting a polygon with 5 faces).

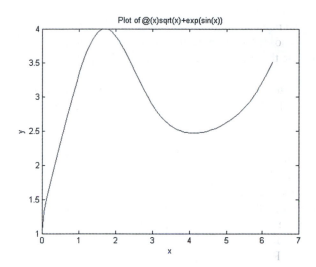

Plot of @(x)sqrt(x)+exp(sin(x))

FIGURE 11.14

Test case for the function `myFunPlotter` `(f,x)` on the function $\sqrt{x} + \exp(\sin(x))$.

.m 8. Write a function with header `[]` = `myFunPlotter` `(f,x)` where f is a function handle and x is an array. The function should plot f evaluated at x. Use the `func2str` and `sprintf` functions to put the function f into the title. Remember to label the x- and y-axis. (See Figure 11.14.)

Test Cases:
```
>> myFunPlotter(@(x) sqrt(x) + exp(sin(x)), linspace(0,2*pi,100))
```

.m 9. Write a function with header `[]` = `myPolyPlotter` `(n,x)` that plots the polynomials $p_k(x) = x^k$ for $k = 1, \ldots, n$. Make sure your plot has axis labels and a title. (See Figure 11.15.)

Test Cases:
```
>> myPolyPlotter(5,-1:.01:1)
```

.m 10. Assume you have three points at the corner of an equilateral triangle, $P_1 = (0,0)$, $P_2 = (0.5, \sqrt{2}/2)$, and $P_3 = (1,0)$. Now you want to generate another set of points $p_i = (x_i, y_i)$ such that $p_1 = (0,0)$ and p_{i+1} is the midpoint between p_i and P_1 with 33% probability, the midpoint between p_i and P_2 with 33% probability, and the midpoint between p_i and P_3 with 33% probability. Write a function with header `[]` = `mySierpinski` `(n)` that generates the points p_i for $i = 1, \cdots, n$. The function should make a plot of the points using blue dots (i.e., `'b.'` as the third argument to the `plot` command). (See Figure 11.16.)

Test Cases:
```
>> mySierpinski(100)
```
Try your function for n = 10000.

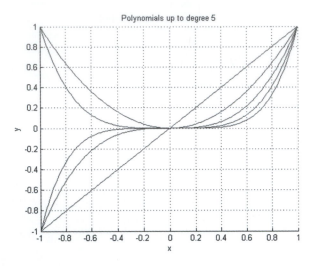

FIGURE 11.15

Test case for the function `myPolyPlotter` `(n,x)` used for five polynomials $p_k(x) = x^1$ for $k = 1, \cdots, 5$.

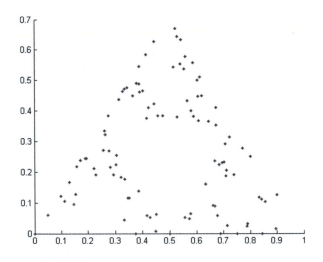

FIGURE 11.16

Test case for the function `mySierpinski` `(n)`.

[m] **11.** Assume you are generating a set of points (x_i, y_i) where $x_1 = 0$ and $y_1 = 0$. The points (x_i, y_i) for $i = 2, \cdots, n$ is generated according to the following probabilistic relationship:

With 1% probability:

$$x_i = 0$$

$y_i = 0.16y_{i-1}$

With 7% probability:

$x_i = 0.2x_{i-1} - 0.26y_{i-1}$

$y_i = 0.23x_{i-1} + 0.22y_{i-1} + 1.6$

With 7% probability:

$x_i = -0.15x_{i-1} + 0.28y_{i-1}$

$y_i = 0.26x_{i-1} + 0.24y_{i-1} + 0.44$

With 85% probability:

$x_i = 0.85x_{i-1} + 0.04y_{i-1}$

$y_i = -0.04x_{i-1} + 0.85y_{i-1} + 1.6$

Write a function with header [] = myFern (n) that generates the points (x_i, y_i) for $i = 1, \ldots, n$ and plots them using blue dots. Hint: Look at the function mySierpinski that you downloaded for Lab 0. (See Figure 11.17.)

Test Cases:

```
>> myFern(100)
```

My Fern with 100 Iterations

FIGURE 11.17

Test case for the function myFern (n) with 100 iterations.

Try your function for n = 10000. The image generated is called a stochastic fractal. Many times it is cheaper (i.e., requires less space) to store the fractal generating code rather than the image. This makes stochastic fractals useful for image compression.

⌊m⌋ **12.** Write a function with header [] = myParametricPlotter (x,y,t) where x and y are handles to the functions x (t) and y (t), respectively, and t is a one-dimensional array.

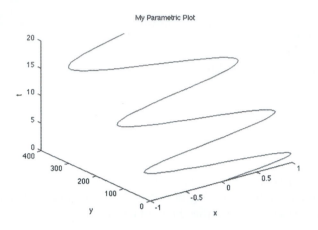

FIGURE 11.18

Test case for `myParametricPlotter`.

The function `myParametricPlotter` should produce the curve $(x(t), y(t), t)$ in a three-dimensional plot. Be sure to give your plot a title and axis labels. (See Figure 11.18.)

Test Cases:

```
>> f = @(t) sin(t);
>> g = @(t) t.^2;
>> myParametricPlotter(f,g,linspace(0,6*pi,100))
```

[m] 13. Write a function with header `[] = mySurfacePlotter (F, x, y, option)` where F is a handle to the function F `(x,y)`. The function `mySurfacePlotter` should produce a 3D surface plot of F `(x,y)` using `surf` if option is the string 'surf'. It should produce a contour plot of F `(x,y)` if the option is the string 'contour'. Assume that x and y are one-dimensional arrays. Remember to give the plot a title and axis labels. (See Figure 11.19.)

Test Cases:

```
>> F = @(X,Y) cos(Y).*sin(exp(X));
>> mySurfacePlotter(F,−1:.1:1, −2:.1:2,'surf')
>> mySurfacePlotter(F,−1:.1:1, −2:.1:2,'contour')
```

14. Write a line of code that generates the following error:

```
??? Error using ==> plot
Vectors must be the same lengths.
```

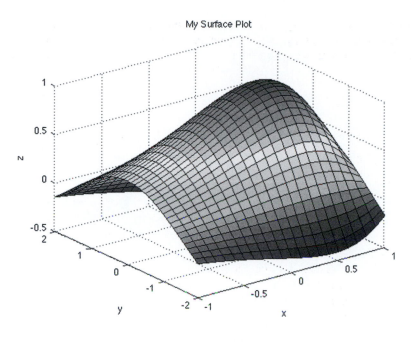

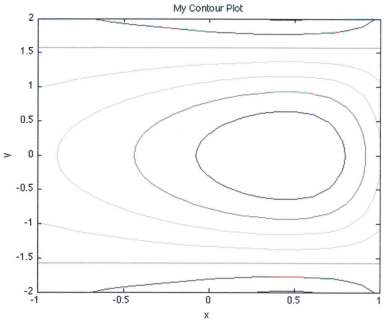

FIGURE 11.19

Test cases for `mySurfacePlotter`.

Introduction to Numerical Methods

PART

II

Introduction
to Numerical
Methods

Linear Algebra and Systems of Linear Equations

Motivation

Numerous problems in engineering can be described or approximated by linear relationships. For example, if you combine resistors in a complicated circuit, you will obtain a system of linear relationships. Similarly, if you study small deformations of rigid structures, you will also get a system of relationships. In fact, it is difficult to think of any technical or engineering field in which relationships of these kind are not fundamental.

The study of linear relationship is contained in the field of linear algebra, and this chapter provides a basic overview of some basic linear algebraic vocabulary and concepts that are important for later chapters. Since this text does not assume any prior knowledge of linear algebra, some of the more abstract mathematical concepts and proofs on this topic have been omitted to make the material more accessible. However, the information in this chapter is in no way comprehensive and should not be considered a substitute for a full linear algebra course.

By the end of this chapter you should understand a variety of linear algebra concepts and calculations. You should know MATLAB's built-in functions for these concepts and calculations. You should know what systems of linear equations are and their relationship to matrices and linear transformations. Finally, you should know how to use MATLAB to compute solutions to systems of linear equations.

12.1 Sets

In mathematics, a **set** is a collection of objects. Sets are usually denoted by braces {}. For example, $S = \{orange, apple, banana\}$ means "S is the set containing 'orange', 'apple', and 'banana'".

An Introduction to MATLAB® Programming and Numerical Methods. http://dx.doi.org/10.1016/B978-0-12-420228-3.00012-9

Table 12.1   Various Sets of Numbers and Corresponding Notations Used to Denote Them

Set Name	Symbol	Description
Naturals	$\mathbb{N}$	$\mathbb{N} = \{1, 2, 3, 4, \cdots\}$.
Wholes	$\mathbb{W}$	$\mathbb{W} = \mathbb{N} \cup \{0\}$
Integers	$\mathbb{Z}$	$\mathbb{Z} = \mathbb{W} \cup \{-1, -2, -3, \cdots\}$
Rationals	$\mathbb{Q}$	$\mathbb{Q} = \{\frac{p}{q} : p \in \mathbb{Z}, q \in \mathbb{Z} \backslash \{0\}\}$
Irrationals	$\mathbb{I}$	$\mathbb{I}$ is the set of real numbers not expressible as a fraction of integers.
Reals	$\mathbb{R}$	$\mathbb{R} = \mathbb{Q} \cup \mathbb{I}$
Complex Numbers	$\mathbb{C}$	$\mathbb{C} = \{a + bi : a, b \in \mathbb{R}, i = \sqrt{-1}\}$

The **empty set** is the set containing no objects and is typically denoted by empty braces such as $\{\}$ or by $\emptyset$. Given two sets, A and B, the **union** of A and B is denoted by $A \cup B$ and equal to the set containing all the elements of A and B. The **intersect** of A and B is denoted by $A \cap B$ and equal to the set containing all the elements that belong to both A and B. In set notation, a colon is used to mean "**such that.**" The usage of these terms will become apparent shortly. The symbol $\in$ is used to denote that an object is contained in a set. For example $a \in A$ means "a is a member of A" or "a is in A." A backslash, \, in set notation means **set minus**. So if $a \in A$ then $A \backslash a$ means "A minus the element, a."

There are several standard sets related to numbers, for example **natural numbers**, **whole numbers**, **integers**, **rational numbers**, **irrational numbers**, **real numbers**, and **complex numbers**. A description of each set and the symbol used to denote them is shown in Table 12.1.

TRY IT! Say whether the following numbers belong to the set of natural numbers, whole numbers, integers, rational numbers, irrational numbers, real numbers, and/or complex numbers: 0, 1, $\pi, e, \sqrt{2}, 3 + 6i$.

$0 : \mathbb{W}, \mathbb{Z}, \mathbb{Q}, \mathbb{R}, \mathbb{C}$.
$1 : \mathbb{N}, \mathbb{W}, \mathbb{Z}, \mathbb{Q}, \mathbb{R}, \mathbb{C}$.
$\pi : \mathbb{I}, \mathbb{R}, \mathbb{C}$.
$e : \mathbb{I}, \mathbb{R}, \mathbb{C}$.
$\sqrt{2} : \mathbb{I}, \mathbb{R}, \mathbb{C}$.
$3 + 6i : \mathbb{C}$.

TRY IT! Let S be the set of all real (x, y) pairs such that $x^2 + y^2 = 1$. Write S using set notation.
$S = \{(x, y) : x, y \in \mathbb{R}, x^2 + y^2 = 1\}$.

12.2 Vectors

The set $\mathbb{R}^n$ is the set of all n-tuples of real numbers. In set notation this is $\mathbb{R}^n = \{(x_1, x_2, x_3, \cdots, x_n) : x_1, x_2, x_3, \cdots, x_n \in \mathbb{R}\}$. For example, the set $\mathbb{R}^3$ represents the set of real triples, (x, y, z) coordinates, in three-dimensional space.

A **vector** in $\mathbb{R}^n$ is an n-tuple, or point, in $\mathbb{R}^n$. Vectors can be written horizontally (i.e., with the elements of the vector next to each other) in a **row vector**, or vertically (i.e., with the elements of the vector on top of each other) in a **column vector**. If the context of a vector is ambiguous, it usually means the vector is a column vector. The i-th element of a vector, v, is denoted by v_i. The transpose of a column vector is a row vector of the same length, and the transpose of a row vector is a column vector. In mathematics, the transpose is denoted by a superscript T, or v^T. In MATLAB, the transpose of a vector, v, is written v'. The **zero vector** is the vector in $\mathbb{R}^n$ containing all zeros.

The **norm** of a vector is a measure of its length. There are many ways of defining the length of a vector depending on the metric used (i.e., the distance formula chosen). The most common is called the L_2 norm, which is computed according to the distance formula you are probably familiar with from grade school. The L_2 **norm** of a vector v is denoted by $\|v\|_2$ and $\|v\|_2 = \sqrt{\sum_i v_i^2}$. This is sometimes also called Euclidian length and refers to the "physical" length of a vector in one-, two-, or three-dimensional space. The L_1 norm, or "Manhattan Distance," is computed as $\|v\|_1 = \sum_i |v_i|$, and is named after the grid-like road structure in New York City. In general, the **p-norm**, L_p, of a vector is $\|v\|_p = \sqrt[p]{(\sum_i v_i^p)}$. The L_∞ **norm** is the p-norm, where $p = \infty$. The L_∞ norm is written as $\|v\|_\infty$ and it is equal to the maximum absolute value in v.

TRY IT! Use the MATLAB function `norm` to compute the L_1, L_2, and L_∞ norm of the vector $v = (1, -5, 3, 2, 4) \in \mathbb{R}^5$. Verify that the L_∞ norm of a vector is equivalent to the maximum value of the elements in the vector.

```
>> v = [1 -5 3 2 4];

>> norm(v, 1)
ans =
    15

>> norm(v, 2)
ans =
    7.4162

>> norm(v,inf)
ans =
    5
>> max(abs(v))
ans =
    5
```

Vector addition is defined as the pairwise addition of each of the elements of the added vectors. For example, if v and w are vectors in $\mathbb{R}^n$, then $u = v + w$ is defined as $u_i = v_i + w_i$.

Vector multiplication can be defined in several ways depending on the context. **Scalar multiplication** of a vector is the product of a vector and a **scalar** (i.e., a number in $\mathbb{R}$). Scalar multiplication is defined as the product of each element of the vector by the scalar. More specifically, if α is a scalar and v is a vector, then $u = \alpha v$ is defined as $u_i = \alpha v_i$. Note that this is exactly how MATLAB implements scalar multiplication with a vector.

TRY IT! Show that $a(v + w) = av + aw$ (i.e., scalar multiplication of a vector distributes across vector addition).

By vector addition, $u = v + w$ is the vector with $u_i = v_i + w_i$. By scalar multiplication of a vector, $x = \alpha u$ is the vector with $x_i = \alpha(v_i + w_i)$. Since α, v_i, and w_i are scalars, multiplication distributes and $x_i = \alpha v_i + \alpha w_i$. Therefore, $a(v + w) = av + aw$.

The **dot product** of two vectors is the sum of the product of the respective elements in each vector and is denoted by $\cdot$, and $v \cdot w$ is read "v dot w." Therefore for v and $w \in \mathbb{R}^n$, $d = v \cdot w$ is defined as $d = \sum_{i=1}^n v_i w_i$. The **angle between two vectors**, θ, is defined by the formula:

$$v \cdot w = \|v\|_2 \|w\|_2 \cos \theta.$$

The dot product is a measure of how similarly directed the two vectors are. For example, the vectors $(1,1)$ and $(2,2)$ are parallel. If you compute the angle between them using the dot product, you will find that $\theta = 0$. If the angle between the vectors, $\theta = \pi/2$, then the vectors are said to be perpendicular or **orthogonal**, and the dot product is 0.

TRY IT! Compute the angle between the vectors $v = (10, 9, 3)$ and $w = (2, 5, 12)$.

```
>> v = [10 9 3];
>> w = [2 5 12];
>> theta = acos(v*w'/(norm(v)*norm(w)))
theta =
    0.9799
```

Note in the present example that the dot product is computed by multiplying v by w's transpose. It could also have been done using MATLAB's function, `dot`.

Finally, the **cross product** between two vectors, v and w, is written $v \times w$. It is defined by $v \times w = \|v\|_2 \|w\|_2 \sin(\theta) n$, where θ is the angle between the v and w (which can be computed from the dot product) and n is a vector perpendicular to both v and w with unit length (i.e., the length is one). The geometric interpretation of the cross product is a vector perpendicular to both v and w with length equal to the area enclosed by the parallelogram created by the two vectors.

TRY IT! Given the vectors $v = (0, 2, 0)$ and $w = (3, 0, 0)$, use the MATLAB function `cross` to compute the cross product of v and w.

```
>> v = [0 2 0];
>> w = [3 0 0];
>> u = cross(v,w)
u =
      0     0    -6
```

Assuming that S is a set in which addition and scalar multiplication are defined, a **linear combination** of S is defined as

$$\sum \alpha_i s_i ,$$

where α_i is any real number and s_i is the ith object in S. Sometimes the α_i values are called the **coefficients** of s_i. Linear combinations can be used to describe numerous things. For example, a grocery bill can be written $\sum c_i n_i$, where c_i is the cost of item i and n_i is the number of item i purchased. Thus, the total cost is a linear combination of the items purchased.

A set is called **linearly independent** if no object in the set can be written as a linear combination of the other objects in the set. For the purposes of this book, we will only consider the linear independence of a set of vectors. A set of vectors that is not linearly independent is **linearly dependent**.

TRY IT! Given the MATLAB vectors v = [0 3 2]', w = [4 1 1]', and u = [0 -2 0]', write the vector x = [-8 -1 4]' as a linear combination of v, w, and u.

```
>> v = [0 3 2]'
>> w = [4 1 1]'
>> u = [0 -2 0]'
>> x = 3*v - 2*w + 4*u
x =
      -8
      -1
       4
```

TRY IT! Determine by inspection whether the following set of vectors is linearly independent: $v = (1, 1, 0)$, $w = (1, 0, 0)$, $u = (0, 0, 1)$.

Clearly u is linearly independent from v and w because only u has a nonzero third element. The vectors v and w are also linearly independent because only v has a nonzero second element. Therefore, v, w, and u are linearly independent.

12.3 Matrices

An $m \times n$ **matrix** is a rectangular table of numbers consisting of m rows and n columns. Matrix addition and scalar multiplication for matrices work the same way as for vectors. However, **matrix multiplication** between two matrices, P and Q, is defined when P is an $m \times p$ matrix and Q is a $p \times n$ matrix. The result of $M = PQ$ is a matrix M that is $m \times n$. The dimension with size p is called the **inner matrix dimension**, and the inner matrix dimensions must match (i.e., the number of columns in P and the number of rows in Q must be the same) for matrix multiplication to be defined. The dimensions m and n are called the **outer matrix dimensions**. Formally, if P is $m \times p$ and Q is $p \times n$, then $M = PQ$ is defined as

$$M_{ij} = \sum_{k=1}^{p} P_{ik} Q_{kj}$$

The product of two matrices P and Q in MATLAB is achieved by the command $*$. The **transpose** of a matrix is a reversal of its rows with its columns. The transpose is denoted by a superscript, T. In MATLAB, the transpose operator is denoted by an apostrophe, '. For example, if M is a matrix, then M' is its transpose.

TRY IT! Let the MATLAB matrices P = [1 7; 2 3; 5 0] and Q = [2 6 3 1; 1 2 3 4]. Compute the matrix product of P and Q using MATLAB's $*$ symbol. Show that Q$*$P and P$*$Q' produce an error.

```
>> P = [1 7; 2 3; 5 0]
Q = [2 6 3 1; 1 2 3 4]
P =
        1       7
        2       3
        5       0
Q =
        2       6       3       1
        1       2       3       4
>> P*Q
ans =
        9      20      24      29
        7      18      15      14
       10      30      15       5
>> P*Q'
??? Error using ==> mtimes
Inner matrix dimensions must agree.

>> Q*P
??? Error using ==> mtimes
Inner matrix dimensions must agree.
```

A **square matrix** is an $n \times n$ matrix; that is, it has the same number of rows as columns. The **determinant** is an important property of square matrices. The determinant is denoted by det, both in mathematics and in MATLAB. Some of the uses of a determinant will be described later.

The **identity matrix** is a square matrix with ones on the diagonal and zeros elsewhere. The identity matrix is usually denoted by I, and is analagous to the real number identity, 1. That is, multiplying any matrix by I (of compatible size) will produce the same matrix.

TRY IT! Use MATLAB to find the determinant of the MATLAB matrix M = [0 2 1 3; 3 2 8 1; 1 0 0 3; 0 3 2 1]. Use the eye function to produce a 4×4 identity matrix, I. Multiply M by I to show that the result is M.

```
>> M = [0 2 1 3; 3 2 8 1; 1 0 0 3; 0 3 2 1]
M =
        0      2      1      3
        3      2      8      1
        1      0      0      3
        0      3      2      1
>> det(M)
ans =
    -38

>> I = eye(4)
I =
        1      0      0      0
        0      1      0      0
        0      0      1      0
        0      0      0      1
>> M*I
ans =
        0      2      1      3
        3      2      8      1
        1      0      0      3
        0      3      2      1
```

The **inverse** of a square matrix M is a matrix of the same size, N, such that $M \cdot N = I$. The inverse of a matrix is analagous to the inverse of real numbers. For example, the inverse of 3 is $\frac{1}{3}$ because $(3)(\frac{1}{3}) = 1$. A matrix is said to be **invertible** if it has an inverse. The inverse of a matrix is unique; that is, for an invertible matrix, there is only one inverse for that matrix. If M is a square matrix, its inverse is denoted by M^{-1} in mathematics, and it can be computed in MATLAB using the function inv.

Recall that 0 has no inverse for multiplication in the real-numbers setting. Similarly, there are matrices that do not have inverses. These matrices are called **singular**. Matrices that do have an inverse are called **nonsingular**.

One way to determine if a matrix is singular is by computing its determinant. If the determinant is 0, then the matrix is singular; if not, the matrix is nonsingular.

TRY IT! The matrix M (in the previous example) has a nonzero determinant. Use MATLAB's inv function to compute the inverse of M. Show that the matrix P = [0 1 0; 0 0 0; 1 0 1] has a determinant value of 0 and therefore has no inverse. Try to compute the inverse anyway.

```
>> inv(M)
ans =
   -1.5789   -0.0789    1.2368    1.1053
   -0.6316   -0.1316    0.3947    0.8421
    0.6842    0.1842   -0.5526   -0.5789
    0.5263    0.0263   -0.0789   -0.3684

>> P = [0 1 0; 0 0 0; 1 0 1]
P =
     0     1     0
     0     0     0
     1     0     1
>> det(P)
ans =
     0
>> inv(P)
ans =
   Inf   Inf   Inf
   Inf   Inf   Inf
   Inf   Inf   Inf
```

A matrix that is close to being singular (i.e., the determinant is close to 0) is called **ill-conditioned**. Although ill-conditioned matrices have inverses, they are problematic numerically in the same way that dividing a number by a very, very small number is problematic. That is, it can result in computations that result in overflow, underflow, or numbers small enough to result in significant round-off errors. The **condition number** is a measure of how ill-conditioned a matrix is, and it can be computed using MATLAB's built-in function cond. The higher the condition number, the closer the matrix is to being singular.

The **rank** of an $m \times n$ matrix A is the number of linearly independent columns or rows of A, and is denoted by rank(A). It can be shown that the number of linearly independent rows is always equal to the number of linearly independent columns for any matrix. A matrix is called **full rank** if rank $(A) = \min(m, n)$. The matrix, A, is also full rank if all of its columns are linearly independent. An **augmented matrix** is a matrix, A, concatenated with a vector, y, and is written $[A, y]$. This is commonly read "A augmented with y." If rank($[A, y]$) = rank(A) + 1, then the vector, y, is "new" information. That is, it cannot be created as a linear combination of the columns in A. The rank is an important property of matrices because of its relationship to solutions of linear equations, which is discussed in the last section of this chapter.

For a square matrix, M, the determinant is zero when M is not full rank; otherwise the determinant is nonzero. As a consequence, you can check if M is singular by checking the determinant.

12.4 Linear Transformations

For vectors x and y, and scalars a and b, it is sufficient to say that a function, F, is a **linear transformation** if

$$F(ax + by) = aF(x) + bF(y).$$

It can be shown that multiplying an $m \times n$ matrix, A, and an $n \times 1$ vector, v, of compatible size is a linear transformation of v. Therefore from this point forward, a matrix will be synonymous with a linear transformation function.

TRY IT! Let x be a vector and let $F(x)$ be defined by $F(x) = Ax$ where A is a rectangular matrix of appropriate size. Show that $F(x)$ is a linear transformation.

Proof:
Since $F(x) = Ax$, then for vectors v and w, and scalars a and b, $F(av + bw) = A(av + bw)$ (by definition of F) $= aAv + bAw$ (by distributive property of matrix multiplication) $= aF(v) + bF(w)$ (by definition of F).
QED.

The **domain** of A is the set of all vectors that can be multiplied by A on the right. If A is an $m \times n$ matrix, then the domain of the linear transformation A is $\mathbb{R}^n$. The **range** of A is the set of all vectors y such that $y = Ax$. Another way to think about the range of A is the set of all linear combinations of the columns in A, where x_i is the coefficient of the ith column in A. The **null space** of A is the subset of vectors in the domain of A, x, such that $Ax = 0$, where $\mathbf{0}$ is the **zero vector** (i.e., a vector in $\mathbb{R}^m$ with all zeros).

TRY IT! Let A = [1 0 0; 0 1 0; 0 0 0] and let the domain of A be $\mathbb{R}^3$. Characterize the range and nullspace of A.
    Let $v = (x, y, z)$ be a vector in $\mathbb{R}^3$. Then $u = Av$ is the vector $u = (x, y, 0)$. Since $x, y \in \mathbb{R}$, the range of A is the x-y plane at $z = 0$.
    Let $v = (0, 0, z)$ for $z \in \mathbb{R}$. Then $u = Av$ is the vector $u = (0, 0, 0)$. Therefore, the nullspace of A is the z-axis (i.e., the set of vectors $(0, 0, z)$ $z \in \mathbb{R}$).
    Therefore, this linear transformation "flattens" any z-component from a vector.

12.5 Systems of Linear Equations

A **linear equation** is an equality of the form

$$\sum_{i=1}^{n} (a_i x_i) = y,$$

where a_i are scalars, x_i are unknown variables in $\mathbb{R}$, and y is a scalar.

TRY IT! Determine which of the following equations is linear and which is not. For the ones that are not linear, can you manipulate them so that they are?

1. $3x_1 + 4x_2 - 3 = -5x_3$
2. $\frac{-x_1+x_2}{x_3} = 2$
3. $x_1x_2 + x_3 = 5$

Equation 1 can be rearranged to be $3x_1 + 4x_2 + 5x_3 = 3$, which clearly has the form of a linear equation. Equation 2 is not linear but can be rearranged to be $-x_1 + x_2 - 2x_3 = 0$, which is linear. Equation 3 is not linear.

A **system of linear equations** is a set of linear equations that share the same variables. Consider the following system of linear equations:

$$
\begin{array}{llll}
a_{1,1}x_1 + & a_{1,2}x_2 + \ldots + & a_{1,n-1}x_{n-1} + & a_{1,n}x_n = & y_1, \\
a_{2,1}x_1 + & a_{2,2}x_2 + \ldots + & a_{2,n-1}x_{n-1} + & a_{2,n}x_n = & y_2, \\
& \cdots & \cdots & \\
a_{m-1,1}x_1 + a_{m-1,2}x_2 + & \ldots + a_{m-1,n-1}x_{n-1} + & a_{m-1,n}x_n = & y_{m-1}, \\
a_{m,1}x_1 + & a_{m,2}x_2 + \ldots + & a_{m,n-1}x_{n-1} + & a_{m,n}x_n = & y_m.
\end{array}
$$

where $a_{i,j}$ and y_i are real numbers. The **matrix form** of a system of linear equations is $Ax = y$ where A is a $m \times n$ matrix, $A(i, j) = a_{i,j}$, y is a vector in $\mathbb{R}^m$, and x is an unknown vector in $\mathbb{R}^n$. If you carry out the matrix multiplication, you will see that you arrive back at the original system of equations.

TRY IT! Put the following system of equations into matrix form.

$$
\begin{aligned}
4x + 3y - 5z &= 2 \\
-2x - 4y + 5z &= 5 \\
7x + 8y &= -3 \\
x + 2z &= 1 \\
9 + y - 6z &= 6
\end{aligned}
$$

$$
\begin{bmatrix}
4 & 3 & -5 \\
-2 & -4 & 5 \\
7 & 8 & 0 \\
1 & 0 & 2 \\
9 & 1 & -6
\end{bmatrix}
\begin{bmatrix} x \\ y \\ z \end{bmatrix}
=
\begin{bmatrix} 2 \\ 5 \\ -3 \\ 1 \\ 6 \end{bmatrix}
$$

12.6 **Solutions to Systems of Linear Equations**

Consider a system of linear equations in matrix form, $Ax = y$, where A is an $m \times n$ matrix. Recall that this means there are m equations and n unknowns in our system. A **solution** to a system of linear equations is an x in $\mathbb{R}^n$ that satisfies the matrix form equation. Depending on the values that populate A and y, there are three distinct solution possibilities for x. Either there is no solution for x, or there is one, unique solution for x, or there are an infinite number of solutions for x. This fact is not shown in this text.

MATLAB has numerous built-in functions that can be used to solve this matrix equation. A useful operator to introduce is the **backslash** or **left-divide** operator, which is denoted by the symbol, \. The command A\y is read, "A left-divides y." The following paragraphs show how MATLAB can be used to find solutions to matrix equations and some rationale behind each of the cases.

WARNING! It is extremely important to note that the left-divide operator can also return a solution for x when there is no solution to the matrix equation. The reason for this is covered in more detail in the next chapter on least-squares regression. Because of operational quirks like this, it is critical to have an understanding of how any solution algorithm you use works.

Case 1: There is no solution for x. If rank($[A, y]$) = rank(A) + 1, then y is linearly independent from the columns of A. Therefore y is not in the range of A and by definition, there cannot be an x that satisfies the equation. Thus, comparing rank($[A, y]$) and rank(A) provides an easy way to check if there are no solutions to a system of linear equations. Avoid using A\y to give you a solution until you understand why you would want to do this in the next chapter.

Case 2: There is a unique solution for x. If rank($[A, y]$) = rank(A), then y can be written as a linear combination of the columns of A and there is at least one solution for the matrix equation. For there to be only one solution, rank(A) = n must also be true. In other words, the number of equations must be exactly equal to the number of unknowns. To see why property results in a unique solution, consider the following three relationships between m and n : $m < n, m = n$, and $m > n$.

1. For the case where $m < n$, rank(A) = n cannot possibly be true because this means we have a "fat" matrix with fewer equations than unknowns. Thus, we do not need to consider this subcase.
2. When $m = n$ and rank(A) = n, then A is square and invertible. Since the inverse of a matrix is unique, then the matrix equation $Ax = y$ can be solved by multiplying each side of the equation, on the left, by A^{-1}. This results in $A^{-1}Ax = A^{-1}y \rightarrow Ix = A^{-1}y \rightarrow x = A^{-1}y$, which gives the unique solution to the equation. This unique solution can be found using x = inv.A/*y or x = A\b.
3. If $m > n$, then there are more equations than unknowns. However if rank(A) = n, then it is possible to choose n equations (i.e., rows of A) such that if these equations are satisfied, then the remaining $m - n$ equations will be also satisfied. In other words, they are redundant. If the

$m - n$ redundant equations are removed from the system, then the resulting system has an A matrix that is $n \times n$, and invertible. These facts are not proven in this text. The new system then has a unique solution, which is valid for the whole system. This unique solution can be found using x = A\b. The command x = inv.A/*y will return an error because A is not square.

Case 3: There is an infinite number of solutions for x. If rank$([A, y]) = $ rank(A), then y is in the range of A, and there is at least one solution for the matrix equation. However, if rank$(A) < n$, then there is an infinite number of solutions. The reason for this fact is as follows: although it is not shown here, if rank$(A) < n$, then there is at least one nonzero vector, n, that is in the null space of A [2]. If n is in the nullspace of A, then $An = 0$ by definition. Now let x^* be a solution to the matrix equation $Ax = y$; then necessarily, $Ax^* = y$. However, $Ax^* + An = y$ or $A(x^* + n) = y$. Therefore, $x^* + n$ is also a solution for $Ax = y$. In fact, since A is a linear transformation, $x^* + \alpha n$ is a solution for any real number, α (you should try to show this on your own). Since there are an infinite number of acceptable values for α, there are an infinite number of solutions for the matrix equation.

You can compute a particular solution to the equation using the command x = pinv.A/*y. Here the function pinv computes the "pseudo inverse" of A, which is discussed in more detail in the next chapter on least squares regression.

The flow chart diagram shown in Figure 12.1 is a good procedure for solving systems of linear equations in MATLAB. However, the solutions returned by MATLAB via this method may include some numerical error due to the estimation procedures employed. While there are numerous factors that contribute to this numerical error, the most important one relates to the general nature of the matrix A. In particular, if A is ill-conditioned, this inversion process will produce significant and problematic numerical errors, the nature of which is beyond of the scope of this book.

EXAMPLE: Illustration of Case 1. We pick a matrix A and a vector y such that y is not in the range of A. Therefore, there is no solution to the matrix equation.

```
>> A = [1,2,3; 0, 3, 1; 1, 14, 7]
A =

         1       2       3
         0       3       1
         1      14       7

>> y = [1; 2; 3]
y =

         1
         2
         3
```

---

[1] Actually there are an infinite number of null space vectors under these conditions.

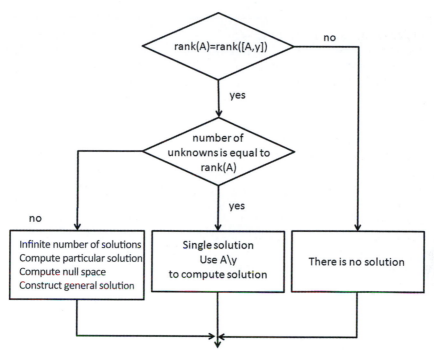

FIGURE 12.1

Illustration of the process for finding the solution to a linear system. There are three scenarios: (1) If y is not in the range of A, there is no solution; (2) If y is in the range of A and the rank of A is equal to the number of unknowns, there is a unique solution; or (3) if y is in the range of A and the rank of A is not equal to the number of unknowns, there is an infinity of solutions.

```
>> rank(A)
ans =
       2

>> rank([A,y])
ans =
       3
```

EXAMPLE: Illustration of Case 2. We pick a matrix A and a vector y such that y is in the range of A, and such that the rank of A is equal to the number of unknowns.

```
>> A = [1,2,3; 0, 3, 1; -1, 14, 7]
A =
        1       2       3
        0       3       1
       -1      14       7

>> y = [1; 2; 3]
y =
        1
        2
        3

>> rank(A)
ans =
        3

>> rank([A y])
ans =
        3

>> x=A\y
x =
    3.0000
    1.1429
   -1.4286

>> x=inv(A)*y
x =
    3.0000
    1.1429
   -1.4286
```

We also give an example with square matrices:

```
>>A = [1,2,3; 0, 3, 1; -1, 14, 7]
A =
        1       2       3
        0       3       1
       -1      14       7

>>y = A*[1;1;1]
y =
        6
        4
       20
```

```
>>x = inv(A)*y
  x =
          1.0000
          1.0000
          1.0000
```

The last command produces the expected result; that is, the result of the inversion of the matrix. It can also be obtained as follows:

```
x = A\y
x =
          1
          1
          1
```

which produces a result with better numerical accuracy. To check that it does what it is supposed to do, we can see that

```
>>A*x
ans =
          6
          4
         20
```

Now if we augment A by one row at the bottom, so it is not square anymore, we get a nonsquare matrix. Let us look at what needs to be done in this case:

```
>>A = [1,2,3; 0, 3, 1; −1, 14, 7; 1 1 1]
A =
          1       2       3
          0       3       1
         −1      14       7
          1       1       1
```

If we reiterate the previous process, we get:

```
>>y= A*[1;1;1]
  y =
          6
          4
         20
          3
```

There is a solution to this system since it was constructed this way (i.e., y was obtained by multiplication of x by A). We can solve for this solution using the commands

```
>>x = A\y
x =
     1.0000
     1.0000
     1.0000
```

Note that using the inverse command would not work in this case because the matrix *A* is not square.

```
>>x=inv(A)*y
??? Error using ==> inv
Matrix must be square.

Error in ==> testOfMatrices2 at 14
x=inv(A)*y
```

EXAMPLE: Illustration of Case 3. We first illustrate this case with a square matrix.

```
A = [1,2,3,0 ; 0, 3, 1 0 ; −1, 14, 7,0; 0 0 0 0]
y = A*[1;1;1;0]
x = pinv(A)*y
A*x
A*[0;0;0;1]
A*(x+[0;0;0;100])
```

The first line defines the following matrix:

```
A =
     1      2      3      0
     0      3      1      0
    −1     14      7      0
     0      0      0      0
```

We can quickly check that the matrix is not invertible (i.e., that its determinant is 0). Looking at the set of commands just above, it is clear that *y* is in the range of *A* since it has been obtained by multiplication of [1;1;1;0] by *A*.

```
y =
     6
     4
    20
     0
```

Now the pinv command computes the pseudo inverse of *A*, which when applied to *y* produces a solution for this equation.

```
x =
     1.0000
     1.0000
     1.0000
          0
```

We can indeed check that A*x returns the expected result.

```
ans =
      6.0000
      4.0000
     20.0000
           0
```

We can also verify that the following command > > A*[0;0;0;1] produces zero; that is, [0;0;0;1] is in the null space of A:

```
ans =
     0
     0
     0
     0
```

Thus, if we add [0;0;0;100] to the previously obtained solution of the system, and then check that it still a solution using the command A*(x+[0;0;0;100]), we get the same result as expected.

```
ans =
      6.0000
      4.0000
     20.0000
           0
```

To investigate the example of a nonsquare matrix, let us consider the following script:

```
% Define A and y
A = [1 2 3 4 5 6; 2 3 4 5 6 7; 3 4 5 6 7 8]
y = [21 ; 27 ; 33]

% Specify an x and verify that it is a solution to Ax = y
x = [1 1 1 1 1 1]'
y = A*x

% Specify a different x and verify that it is a solution
x = [3 0 0 0 0 3]'
y = A*x
```

```
% generate a solution using pinv
x = pinv(A)*y

% generate vectors in nullspace using null function
N = null(A)
n1=N(:,1); n2=N(:,2); n3=N(:,3); n4=N(:,4);

% verify that a linear combination of solutions and null space vectors
% is still a solution
y = A*(x+31*n1+202*n2+87*n3+42*n4)
```

It will produce the following results:

```
A =

        1       2       3       4       5       6
        2       3       4       5       6       7
        3       4       5       6       7       8

y =

       21
       27
       33

x =

        1
        1
        1
        1
        1
        1

y =

       21
       27
       33
```

At this stage, we just checked that this x is a solution to the problem. Let us now look at the following x:

```
x =

        3
        0
        0
        0
        0
        3
```

```
y =
    21
    27
    33
```

This value for x is also a solution to the problem; therefore, there are an infinite number of solutions. If we find a solution for x using `pinv`, it will return the following:

```
x =
    1.0000
    1.0000
    1.0000
    1.0000
    1.0000
    1.0000
```

which is a good numerical approximation of the x above. Now we continue to execute the script and compute vectors in the nullspace.

```
N =
     0.6742   -0.1035    0.1009    0.0266
    -0.6142    0.0243    0.2136    0.5304
    -0.3019   -0.3701   -0.5933   -0.4888
     0.0214    0.8729   -0.1404   -0.1920
    -0.0514   -0.2148    0.7018   -0.4042
     0.2719   -0.2087   -0.2827    0.5280
```

As expected, when multiplying the vector x+31*n1+202*n2+87*n3+42*n4 by A, we obtain the same result as if we were just computing Ax, with some small numerical error:

```
y =
    21.0000
    27.0000
    33.0000
```

Summary

1. Linear algebra is the foundation of many engineering fields.
2. Vectors can be considered as points in $\mathbb{R}^n$; addition and multiplication are defined on them, although not necessarily the same as for scalars.
3. A set of vectors is linearly independent if none of the vectors can be written as a linear combination of the others.
4. Matrices are tables of numbers. They have several important properties including the determinant, rank, and inverse.

5. A system of linear equations can be represented by the matrix equation $Ax = y$.

6. The number of solutions to a system of linear equations is related to the rank(A) and the rank($[A, y]$). It can be zero, one, or infinity.

Vocabulary

angle between vectors	irrational number	range
augmented matrix	L_2 norm	rank
backslash	L_∞ norm	rational number
coefficient	left division	real number
column vector	linear combination	row vector
complex number	linear equation	scalar
condition number	linear transformation	scalar multiplication
cross product	linearly dependent	set
determinant	linearly independent	singular
domain	matrix	solution
dot product	matrix form	square matrix
empty set	matrix multiplication	such that
full rank	minus	system of linear equations
identity matrix	natural number	transpose
ill conditioned	nonsingular	union
inner matrix dimensions	norm	vector
integer	nullspace	vector addition
intersect	orthogonal	whole number
inverse	outer matrix dimensions	zero vector
invertible	p-norm	

Functions and Operators

\	inv	range
cond	null	rank
det	pinv	

Problems

1. Show that matrix multiplication distributes over matrix addition: show $A(B + C) = AB + AC$ assuming that A, B, and C are matrices of compatible size.

2. Write a function with header [out] = myIsOrthogonal(v1,v2, tol), where v1 and v2 are column vectors of the same size and tol is a scalar value strictly larger than 0. The output argument, out, should be 1 if the angle between v1 and v2 is within tol of $\pi/2$; that is,

$|\pi/2 - \theta| <$ tol, and 0 otherwise. You may assume that v1 and v2 are column vectors of the same size, and that tol is a positive scalar.

Test Cases:

```
>> out = myIsOrthogonal([1;0.001],[0.001;1],0.01)
out =
     1
>> out = myIsOrthogonal([1;0.001],[0.001;1],0.001)
out =
     0
>> out = myIsOrthogonal([1;0.001],[1;1],0.01)
out =
     0
>> out = myIsOrthogonal([1; 1], [-1; 1], 1e-10)
out =
     1
>> out = myIsOrthogonal([1 0 1 -1 1 1]', [0 1 0 1 1 0]', 1e-10)
out =
     1
```

⌘ 3. Write a function with header [out] = myIsSimilar(s1,s2,tol) where s1 and s2 are strings, not necessarily the same size, and tol is a scalar value strictly larger than 0. From s1 and s2, myIsSimilar should construct two vectors, v1 and v2, where v1(1) is the number of 'a's in s1, v1(2) is the number 'b's in s1, and so on until v1(26), which is the number of 'z's in v1. The vector v2 should be similarly constructed from s2. The output argument, out, should be 1 if the absolute value of the angle between v1 and v2 is less than tol; that is, $|\theta| <$ tol.

⌘ 4. Write a function with header [B] = myMakeLinInd(A), where A and B are matrices. Let the rank(A) = n. Then B should be a matrix containing the first n columns of A that are all linearly independent. Note that this implies that B is full rank.

Test Cases:

```
>> A = [
    12    24     0    11   -24    18    15    14    -2
    19    38     0    10   -31    25     9     4    11
     1     2     0    21    -5     3    20    24    -7
     6    12     0    13   -10     8     5     7     4
    22    44     0     2   -12    17    23    16    -7];
>> B = myMakeLinInd(A)
B =
    12    11   -24    15    14
    19    10   -31     9     4
     1    21    -5    20    24
     6    13   -10     5     7
    22     2   -12    23    16
```

⌞m 5. Cramer's rule is a method of computing the determinant of a matrix. Consider an $n \times n$ square matrix M. Let $M(i, j)$ be the element of M in the i-th row and j-th column of M, and let $m_{i,j}$ be the minor of M created by removing the i-th row and j-th column from M. Cramer's rule says that

$$\det(M) = \sum_{i=1}^{n} (-1)^{i-1} M(1, i)\det(m_{i,j}).$$

Write a function with header $[D]$ = myRecDet(M), where D is det(M). myRecDet should use Cramer's rule to compute the determinant, not MATLAB's function det.

6. What is the complexity of myRecDet in the previous problem? Do you think this is an effective way of determining if a matrix is singular or not?

⌞m 7. Let p be a vector with length L containing the coefficients of a polynomial of order L−1. For example, the vector p = [1; 0; 2] is a representation of the polynomial $f(x) = 1x^2 + 0x + 2$. Write a function with header $[D]$ = myPolyDerMat(p), where p is the aforementioned vector, and D is the matrix that will return the coefficients of the derivative of p when p is left multiplied by D. For example, the derivative of $f(x)$ is $f'(x) = 2x$, and therefore, $d = Dp$ should yield d = [2; 0]. Note this implies that the dimension of D is $L − 1 \times L$. The point of this problem is to show that integrating polynomials is actually a linear transformation.

⌞m 8. Write a function with header $[N, x]$ = myNumSols(A,b), where A and b are a matrix and compatibly-sized column vector, respectively; N is the number of solutions of the system $Ax = b$; and x is a solution to the same system. If there are no solutions to the system of equations, then x should be an empty matrix. If there is one or an infinite number of solutions, then myNumSols should return one using the methods described in the chapter. You may assume that b is a column vector and that the number of elements in b is the same as the number of rows in A. The output x should be a column vector. You may assume that if the system has an infinite number of solutions, then the A matrix will have more columns than rows; that is, A is fat. In this case, you should solve the system using x = pinv(A)*b.

Test Cases:

```
>> A = reshape(1:15, 3, 5);
>> b = [−5; −4; −3];
>> [N, x] = myNumSols(A,b)
N =
    Inf
x =
    1.0000
    0.6000
    0.2000
   −0.2000
   −0.6000

>> b = [−1.5; 2; 7];
>> [N, x] = myNumSols(A,b)
```

```
N =

      0
x =

     [ ]

>> A = 3*eye(5);
>>b = [1; 2; 3; 4; 5];
>> [N, x] = myNumSols(A,b)
N =

     1
x =
     0.3333
     0.6667
     1.0000
     1.3333
     1.6667
```

.m 9. Consider the following network consisting of two power supply stations denoted by S1 and S2 and five power recipient nodes denoted by N1 to N5. The nodes are connected by power lines, which are denoted by arrows, and power can flow between nodes along these lines in both directions.

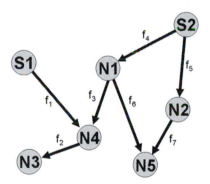

Let d_i be a positive scalar denoting the power demands for node i, and assume that this demand must be met exactly. The capacity of the power supply stations is denoted by S. Power supply stations must run at their capacity. For each arrow, let f_j be the power flow along that arrow. Negative flow implies that power is running in the opposite direction of the arrow.

Write a function with header [f] = myFlowCalculator(S, d), where S is a 1×2 vector representing the capacity of each power supply station, and d is a 1×5 row vector representing the demands at each node (i.e., d(1) is the demand at node 1). The output argument, f, should be a 1×7 row vector denoting the flows in the network (i.e., $f(1) = f_1$ in the diagram). The flows contained in f should satisfy all constraints of the system, like power generation and demands. Note that there may be more than one solution to the system of equations.

The total flow into a node must equal the total flow out of the node plus the demand; that is, for each node i, $f_{inflow} = f_{outflow} + d_i$. You may assume that $\Sigma S_j = \Sigma d_i$.

Test Cases:

```
>> f = myFlowCalculator([10 10], [4 4 4 4 4])
f =
    10.0000    4.0000    -2.0000    4.5000    5.5000    2.5000    1.5000

>> f = myFlowCalculator([10 10], [3 4 5 4 4])
f =
    10.0000    5.0000    -1.0000    4.5000    5.5000    2.5000    1.5000
```

10. Show that the dot product distributes across vector addition; that is, show that $u \cdot (v + w) = u \cdot v + u \cdot w$.

Least Squares Regression

13

CHAPTER OUTLINE

Motivation

Often in physics and engineering coursework, we are asked to determine the state of a system given the parameters of the system. For example, the relationship between the force exerted by a linear spring, F, and the displacement of the spring from its natural length, x, is usually represented by the model

$$F = kx,$$

where k is the spring stiffness. We are then asked to compute the force for a given k and x value. However in practice, the stiffness and in general, most of the parameters of a system, are not known *a priori*. Instead, we are usually presented with data points about how the system has behaved in the past. For our spring example, we may be given (x, F) data pairs that have been previously recorded from an experiment. Ideally, all these data points would lie exactly on a line going through the origin (since there is no force at zero displacement). We could then measure the slope of this line and get our stiffness value for k. However, practical data usually has some measurement noise because of sensor inaccuracy, measurement error, or a variety of other reasons. Figure 13.1 shows an example of what data might look like for a simple spring experiment.

This chapter teaches methods of finding the "most likely" model parameters given a set of data; for example, how to find the spring stiffness in our mock experiment. By the end of this chapter you should understand how these methods choose model parameters, the importance of choosing the correct model, and how to implement these methods in MATLAB.

An Introduction to MATLAB® Programming and Numerical Methods. http://dx.doi.org/10.1016/B978-0-12-420228-3.00013-0

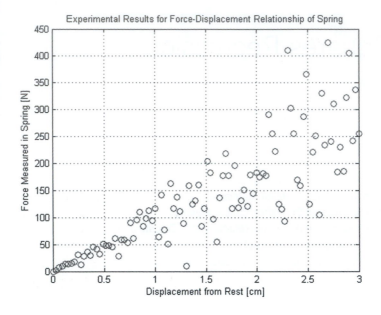

FIGURE 13.1

Results from force-displacement experiment for spring (fictional). The theoretical linear relationship between force and displacement in a linear spring is $F = kx$. What do you think k should be given as the experimental data?

13.1 Least Squares Regression Problem Statement

Given a set of independent data points x_i and dependent data points y_i, $i = 1, \ldots, m$, we would like to find an **estimation function**, $\hat{y}(x)$, that describes the data as well as possible. Note that $\hat{y}$ can be a function of several variables, but for the sake of this discussion, we restrict the domain of $\hat{y}$ to be a single variable. In least squares regression, the estimation function must be a linear combination of **basis functions**, $f_i(x)$. That is, the estimation function must be of the form

$$\hat{y}(x) = \sum_{i=1}^{n} \alpha_i f_i(x)$$

The scalars α_i are referred to as the **parameters** of the estimation function, and each basis function must be linearly independent from the others. In other words, in the proper "functional space" no basis function should be expressible as a linear combination of the other functions. Note: In general, there are significantly more data points, m, than basis functions, n (i.e., $m >> n$).

> **TRY IT!** Create an estimation function for the force-displacement relationship of a linear spring. Identify the basis function(s) and model parameters.

> The relationship between the force, F, and the displacement, x, can be described by the function $F(x) = kx$ where k is the spring stiffness. The only basis function is the function $f_1(x) = x$ and the model parameter to find is $\alpha_1 = k$.

The goal of **least squares regression** is to find the parameters of the estimation function that minimize the **total squared error**, E, defined by $E = \sum_{i=1}^{m} (\hat{y} - y_i)^2$. The **individual errors** or **residuals** are defined as $e_i = (\hat{y} - y_i)$. If e is the vector containing all the individual errors, then we are also trying to minimize $E = \|e\|_2^2$, which is the L_2 norm defined in the previous chapter.

In the next two sections we derive the least squares method of finding the desired parameters. The first derivation comes from linear algebra, and the second derivation comes from multivariable calculus. Although they are different derivations, they lead to the same least squares formula. You are free to focus on the section with which you are most comfortable.

13.2 Least Squares Regression Derivation (Linear Algebra)

First, we enumerate the estimation of the data at each data point x_i.

$$\hat{y}(x_1) = \alpha_1 f_1(x_1) + \alpha_2 f_2(x_1) + \cdots + \alpha_n f_n(x_1),$$
$$\hat{y}(x_2) = \alpha_1 f_1(x_2) + \alpha_2 f_2(x_2) + \cdots + \alpha_n f_n(x_2),$$
$$\cdots$$
$$\hat{y}(x_m) = \alpha_1 f_1(x_m) + \alpha_2 f_2(x_m) + \cdots + \alpha_n f_n(x_m).$$

Let $X \in \mathbb{R}^n$ be a column vector such that the i-th element of X contains the value of the i-th x-data point, x_i, $\hat{Y}$ be a column vector with elements, $\hat{Y}_i = \hat{y}(x_i)$, β be a column vector such that $\beta_i = \alpha_i$, $F_i(x)$ be a function that returns a column vector of $f_i(x)$ computed on every element of x, and A be an $m \times n$ matrix such that the i-th column of A is $F_i(x)$. Given this notation, the previous system of equations becomes $\hat{Y} = A\beta$.

Now if Y is a column vector such that $Y_i = y_i$, the total squared error is given by $E = \|\hat{Y} - Y\|_2^2$. You can verify this by substituting the definition of the L_2 norm. Since we want to make E as small as possible and norms are a measure of distance, this previous expression is equivalent to saying that we want $\hat{Y}$ and Y to be a "close as possible." Note that in general Y will not be in the range of A and therefore $E > 0$.

Consider the following simplified depiction of the range of A; see Figure 13.2. Note this is *not* a plot of the data points (x_i, y_i).

From observation, the vector in the range of A, $\hat{Y}$, that is closest to Y is the one that can point perpendicularly to Y. Therefore, we want a vector $Y - \hat{Y}$ that is perpendicular to the vector $\hat{Y}$.

Recall from the chapter on Linear Algebra that two vectors are perpendicular, or orthogonal, if their dot product is 0. Noting that the dot product between two vectors, v and w, can be written as $\text{dot}(v, w) = v^T w$, we can state that $\hat{Y}$ and $Y - \hat{Y}$ are perpendicular if $\text{dot}(\hat{Y}, Y - \hat{Y}) = 0$; therefore, $\hat{Y}^T(Y - \hat{Y}) = 0$, which is equivalent to $(A\beta)^T(Y - A\beta) = 0$.

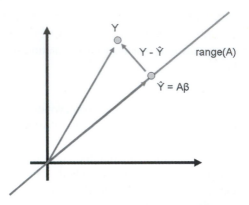

FIGURE 13.2

Illustration of the L_2 projection of Y on the range of A.

Noting that for two matrices A and B, $(AB)^T = B^T A^T$ and using distributive properties of vector multiplication, this is equivalent to $\beta^T A^T Y - \beta^T A^T A\beta = \beta^T (A^T Y - A^T A\beta) = 0$. The solution, $\beta = \mathbf{0}$, is a trivial solution, so we use $A^T Y - A^T A\beta = 0$ to find a more interesting solution. Solving this equation for β gives the **least squares regression formula**:

$$\beta = (A^T A)^{-1} A^T Y$$

Note that $(A^T A)^{-1} A^T$ is called the **pseudo-inverse** of A and exists when $m > n$ and A has linearly independent columns. Proving the invertibility of $(A^T A)$ is outside the scope of this book, but it is always invertible except for some pathological cases.

13.3 Least Squares Regression Derivation (Multivariable Calculus)

Recall that the total error for m data points and n basis functions is:

$$E = \sum_{i=1}^{m} e_i^2 = \sum_{i=1}^{m} (\hat{y}(x_i) - y_i)^2 = \sum_{i=1}^{m} \left(\sum_{j=1}^{n} \alpha_j f_j(x_i) - y_i \right)^2 .$$

which is an n-dimensional paraboloid in α_k. From calculus, we know that the minimum of a paraboloid is where all the partial derivatives equal zero. So taking partial derivative of E with respect to the variable α_k (remember that in this case the parameters are our variables), setting the system of equations equal to 0 and solving for the α_k's should give the correct results.

The partial derivative with respect to α_k and setting equal to 0 yields:

$$\frac{\partial E}{\partial \alpha_k} = \sum_{i=1}^{m} 2 \left(\sum_{j=1}^{n} \alpha_j f_j(x_i) - y_i \right) f_k(x_i) = 0.$$

With some rearrangement, the previous expression can be manipulated to the following:

$$\sum_{i=1}^{m}\sum_{j=1}^{n}\alpha_j f_j(x_i) f_k(x_i) - \sum_{i=1}^{m} y_i f_k(x_i) = 0,$$

and further rearrangement taking advantage of the fact that addition commutes results in:

$$\sum_{j=1}^{n}\alpha_j \sum_{i=1}^{m} f_j(x_i) f_k(x_i) = \sum_{i=1}^{m} y_i f_k(x_i).$$

Now let X be a column vector such that the i-th element of X is x_i and Y similarly constructed, and let $F_j(X)$ be a column vector such that the i-th element of $F_j(X)$ is $f_j(x_i)$. Using this notation, the previous expression can be rewritten in vector notation as:

$$\left[F_k^T(X)F_1(X), F_k^T(X)F_2(X), \ldots, F_k^T(X)F_j(X), \ldots, F_k^T(X)F_n(X) \right] \begin{bmatrix} \alpha_1 \\ \alpha_2 \\ \ldots \\ \alpha_j \\ \ldots \\ \alpha_n \end{bmatrix} = F_k^T(X)Y.$$

If we repeat this equation for every k, we get the following system of linear equations in matrix form:

$$\begin{bmatrix} F_1^T(X)F_1(X), F_1^T(X)F_2(X), \ldots, F_1^T(X)F_j(X), \ldots, F_1^T(X)F_n(X) \\ F_2^T(X)F_1(X), F_2^T(X)F_2(X), \ldots, F_2^T(X)F_j(X), \ldots, F_2^T(X)F_n(X) \\ \ldots \ldots \\ F_n^T(X)F_1(X), F_n^T(X)F_2(X), \ldots, F_n^T(X)F_j(X), \ldots, F_n^T(X)F_n(X) \end{bmatrix} \begin{bmatrix} \alpha_1 \\ \alpha_2 \\ \ldots \\ \alpha_j \\ \ldots \\ \alpha_n \end{bmatrix} = \begin{bmatrix} F_1^T(X)Y \\ F_2^T(X)Y \\ \ldots \\ F_n^T(X)Y \end{bmatrix}.$$

If we let $A = [F_1(X), F_2(X), \ldots, F_j(X), \ldots, F_n(X)]$ and β be a column vector such that j-th element of β is α_j, then the previous system of equations becomes

$$A^T A\beta = A^T Y,$$

and solving this matrix equation for β gives $\beta = (A^T A)^{-1}A^T Y$, which is exactly the same formula as the previous derivation.

13.4 Least Squares Regression in MATLAB®

Recall that if we enumerate the estimation of the data at each data point, x_i, this gives us the following system of equations:

$$\hat{y}(x_1) = \alpha_1 f_1(x_1) + \alpha_2 f_2(x_1) + \cdots + \alpha_n f_n(x_1),$$
$$\hat{y}(x_2) = \alpha_1 f_1(x_2) + \alpha_2 f_2(x_2) + \cdots + \alpha_n f_n(x_2),$$
$$\ldots$$
$$\hat{y}(x_m) = \alpha_1 f_1(x_m) + \alpha_2 f_2(x_m) + \cdots + \alpha_n f_n(x_m).$$

If the data was absolutely perfect (i.e., no noise), then the estimation function would go through all the data points, resulting in the following system of equations:

$$y_1 = \alpha_1 f_1(x_1) + \alpha_2 f_2(x_1) + \cdots + \alpha_n f_n(x_1),$$
$$y_2 = \alpha_1 f_1(x_2) + \alpha_2 f_2(x_2) + \cdots + \alpha_n f_n(x_2),$$
$$\cdots$$
$$y_m = \alpha_1 f_1(x_m) + \alpha_2 f_2(x_m) + \cdots + \alpha_n f_n(x_m).$$

If we take A to be as defined previously, this would result in the matrix equation

$$y = Ax.$$

However, since the data is not perfect, there will not be an estimation function that can go through all the data points, and this system will have *no solution*. However, recall in the previous chapter that x = A\y would return a solution even if no solution existed. It turns out the solution returned by MATLAB for this command is the least squares solution derived in the previous two sections. In other words, if there is a solution, x = A\y will return one, otherwise it will return the x that is the closest to being a solution to the matrix equation.

The pseudo-inverse for of A can be computed using the MATLAB function pinv, which you have already used in the previous chapter to solve systems of linear equations.

TRY IT! For the matrix A = [1 2; 3 4; 5 6] and the vector y = [4; 1; 2], show that x = inv (A'*A)*A'*y, x = pinv (A)*y, and x = A\y all produce the same result for x.

```
>> A = [1 2; 3 4; 5 6];
>> y = [4; 1; 2];

>> x = inv(A'*A)*A'*y
x =
    -4.3333
     3.8333

>> x = pinv(A)*y
x =
    -4.3333
     3.8333

>> x = A\y
x =
    -4.3333
     3.8333
```

TRY IT! Consider the artificial data created by x = 0:.01:1 and y = 1 + x + x.*rand (size (x));. Do a least squares regression with an estimation function defined by

$\hat{y}(x) = \alpha_1 x + \alpha_2$. Plot the data points along with the least squares regression line, as shown in the Figure 13.3. Note that we expect $\alpha_1 = 1$ and $\alpha_2 = 1.5$ based on this data.

```
>> x = [0:.01:1]';
>> y = 1 + x + x.*rand(size(x));
>> A = [x, ones(size(x))];
% the column of ones comes from x^0,
% the basis function associated with alpha(1)
>> alpha = inv(A'*A)*A'*y
alpha =
      1.6111
      0.9686
>> hold on
>> plot(x, alpha(1)*x + alpha(2),'r')
>> plot(x,y,'bo')
>> xlabel('x')
>> ylabel('y')
```

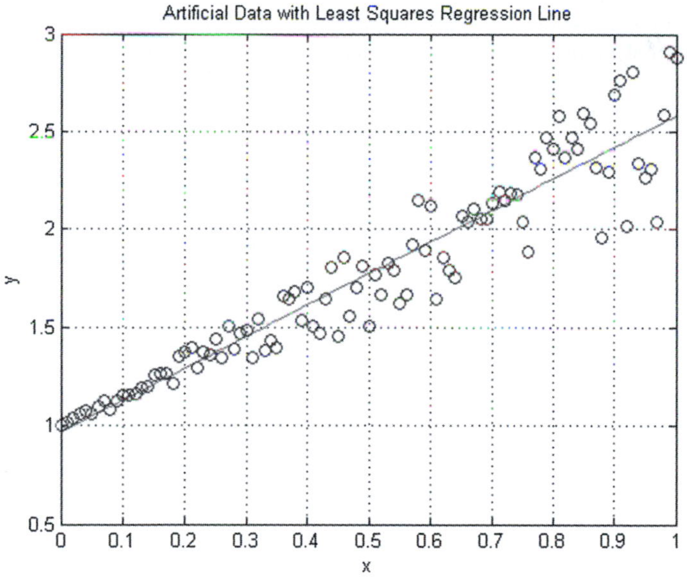

FIGURE 13.3

Plotting resulting from execution of previous code. Estimation data and regression curve $\hat{y}(x) = \alpha_1 x + \alpha_2$.

13.5 Log Tricks for Nonlinear Estimation Functions

Least squares regression requires that the estimation function be a linear combination of basis functions. However, there are some functions that cannot be put in this form but where a least squares regression is still appropriate. We can accomplish this by taking advantage of the properties of logarithms.

Assume you have a function of the form $\hat{y}(x) = \alpha e^{\beta x}$ and data for x and y, and that you want to do least squares regression to find α and β. Clearly, the previous set of basis functions (linear) would be an inappropriate choice to describe $\hat{y}(x)$. However, if we take the log of both sides, we get $\log(\hat{y}(x)) = \log(\alpha) + \beta x$. Now if we say that $\tilde{y}(x) = \log(\hat{y}(x))$ and $\tilde{\alpha} = \log(\alpha)$, then we get $\tilde{y}(x) = \tilde{\alpha} + \beta x$. We can perform least squares regression on the linearized expression to find $\tilde{y}(x), \tilde{\alpha}$, and β, and then recover α by using the expression $\alpha = e^{\tilde{\alpha}}$.

Summary

1. Mathematical models are used to understand, predict, and control engineering systems. These models consist of parameters that govern the way the model behaves.
2. Given a set of experimental data, least squares regression is a method of finding a set of model parameters that fits the data well. That is, it minimizes the squared error between the model, or estimation function, and the data points.
3. In least squares regression, the estimation function must be a linear combination of linearly independent basis functions.
4. The set of parameters β can be determined by the least squares equation $\beta = (A^T A)^{-1} A^T Y$, where the j-th column of A is the j-th basis function evaluated at each X data point.

Vocabulary

basis function	parameters	residuals
estimation function	pseudo-inverse	total squared error
least squares regression		

Functions and Operators

\	polyfit
pinv	polyval

Problems

1. Repeat the multivariable calculus derivation of the least squares regression formula for an estimation function $\hat{y}(x) = ax^2 + bx + c$ where $a, b,$ and c are the parameters.
2. Write a function with header [Beta] = myLSParams (f, x, y), where x and y are column vectors of the same size containing experimental data, and f is a cell array with each element a function handle to a basis vector of the estimation function. The output argument, Beta, should be a column vector of the parameters of the least squares regression for x, y, and f.
3. Write a function with header [alpha, beta] = myExpFit (x,y), where x and y are column vectors of the same size containing experimental data, and alpha and beta are the parameters of the estimation function $\hat{y}(x) = \alpha e^{\beta x}$.

4. Given four data points (x_i, y_i) and the parameters for a cubic polynomial $\hat{y}(x) = ax^3 + bx^2 + cx + d$, what will be the total error associated with the estimation function $\hat{y}(x)$? Where can we place another data point (x, y) such that no additional error is incurred for the estimation function?

5. Write a function with header [beta] = myLinRegression (f, x, y), where f is a cell array containing function handles to basis functions, and x and y are column vectors containing noisy data. Assume that x and y are the same length, and that the functions contained in f are vectorized.

Let an estimation function for the data contained in x and y be defined as $\hat{y}(x) = \beta(1) \cdot f_1(x) + \beta(2) \cdot f_2(x) + \cdots + \beta(n) \cdot f_n(x)$, where n is the length of f. Your function should compute beta according to the least squares regression formula.

Test Case: Note that your solution may vary by a little bit depending on the random numbers generated.

```
>> x = linspace(0,2*pi,1000)';
>> y = 3*sin(x) - 2*cos(x) + randn(size(x));
>> f = {@sin, @cos};
>> beta = myLinRegression(f, x, y)
beta =
     3.0337
    -2.0147
>> plot(x,y,'b.', x, beta(1)*f{1}(x) + beta(2)*f{2}(x),'r', 'LineWidth', 3)
>> title('Least Squares Regression Example')
>> xlabel('x')
>> ylabel('y')
>> legend('data', 'regression')
```

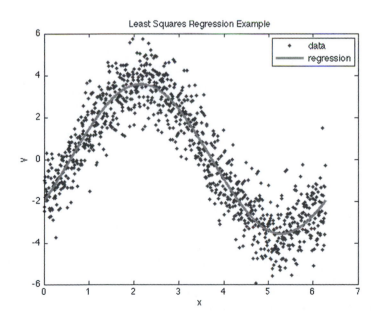

⌷ 6. Write a function with header [alpha, beta] = myExpRegression (x,y), where x
and y are column vectors of the same size.

Let an estimation function for the data contained in x and y be defined as $\hat{y}(x) = \alpha e^{\beta x}$.
Your function should compute α and β as the solution to the least squares regression formula.
To accomplish this, you should first linearize the estimation function according to the methods
specified in Section 13.5.

Test Cases: Note that your solution may vary from the test case slightly depending on the
random numbers generated.

```
>> x = linspace(0,1,1000)';
>> y = 2*exp(-.5*x) + .25*randn(size(x));
>> [alpha, beta] = myExpRegression(x,y)
alpha =
     1.9886
beta =
    -0.5197
>> plot(x,y, 'b.', x, alpha*exp(beta*x), 'r', 'LineWidth', 3)
>> title('Least Squares Regression on Exponential Model')
>> xlabel('x')
>> ylabel('y')
>> legend('data', 'regression')
```

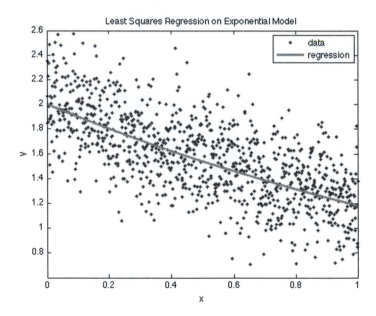

Interpolation

14

CHAPTER OUTLINE

Motivation

The previous chapter used regression to find the parameters of a function that best estimated a set of data points. Regression assumes that the data set has measurement errors, and that you need to find a set of model parameters that minimize the error between your model and the data. However, sometimes you have measurements that are assumed to be very reliable; in these cases, you want an estimation function that goes *through* the data points you have. This technique is commonly referred to as interpolation.

By the end of the chapter, you should be able to understand and compute some of those most common interpolating functions.

14.1 Interpolation Problem Statement

Assume we have a data set consisting of independent data values, x_i, and dependent data values, y_i, where $i = 1, \ldots, n$. We would like to find an estimation function $\hat{y}(x)$ such that $\hat{y}(x_i) = y_i$ for every point in our data set. This means the estimation function goes through our data points. Given a new $x*$, we can **interpolate** its function value using $\hat{y}(x*)$. In this context, $\hat{y}(x)$ is called an **interpolation function**. Figure 14.1 shows the interpolation problem statement.

Unlike regression, interpolation does not require the user to have an underlying model for the data, especially when there are many reliable data points. However, the processes that underly the data must still inform the user about the quality of the interpolation. For example, our data may consist of (x, y) coordinates of a car over time. Since motion is restricted to the maneuvering physics of the car, we can expect that the points between the (x, y) coordinates in our set will be "smooth" rather than jagged.

In the following sections we derive several common interpolation methods.

An Introduction to MATLAB® Programming and Numerical Methods. http://dx.doi.org/10.1016/B978-0-12-420228-3.00014-2
© 2015 Elsevier Inc. All rights reserved.

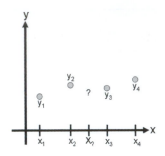

FIGURE 14.1

Illustration of the interpolation problem: estimate the value of a function in between data points.

14.2 Linear Interpolation

In **linear interpolation**, the estimated point is assumed to lie on the line joining the nearest points to the left and right. Assume, without loss of generality, that the x-data points are in ascending order; that is, $x_i < x_{i+1}$, and let x be a point such that $x_i < x < x_{i+1}$. Then the linear interpolation at x is

$$\hat{y}(x) = y_i + \frac{(y_{i+1} - y_i)(x - x_i)}{(x_{i+1} - x_i)}.$$

TRY IT! Find the linear interpolation at x = 1.5 based on the data x = [0 1 2], y = [1 3 2]. Verify the result using MATLAB's function interp1. (See Figure 14.2.)

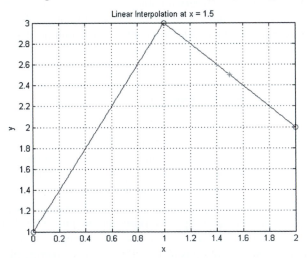

FIGURE 14.2

Linear interpolation of the points $x = (0, 1, 2)$ and $y = (1, 3, 2)$.

Since $1 < x < 2$, we use the second and third data points to compute the linear interpolation. Plugging in the corresponding values gives

$$\hat{y}(x) = y_i + \frac{(y_{i+1} - y_i)(x - x_i)}{(x_{i+1} - x_i)} = 3 + \frac{(2-3)(1.5-1)}{(2-1)} = 2.5$$

```
>> yhat = interp1([0 1 2], [1 3 2], 1.5)
yhat =
    2.5000
```

14.3 Cubic Spline Interpolation

In **cubic spline interpolation** (Figure 14.3), the interpolating function is a set of piecewise cubic functions. Specifically, we assume that the points (x_i, y_i) and (x_{i+1}, y_{i+1}) are joined by a cubic polynomial $S_i(x) = a_i x^3 + b_i x^2 + c_i x + d_i$ that is valid for $x_i \le x \le x_{i+1}$ for $i = 1, \ldots, n-1$. To find the interpolating function, we must first determine the coefficients a_i, b_i, c_i, d_i for each of the cubic functions. For n points, there are $n-1$ cubic functions to find, and each cubic function requires four coefficients. Therefore we have a total of $4(n-1)$ unknowns, and so we need $4(n-1)$ independent equations to find all the coefficients.

First we know that the cubic functions must intersect the data the points on the left and the right:

$$S_i(x_i) = y_i, \quad i = 1, \ldots, n-1,$$
$$S_i(x_{i+1}) = y_{i+1}, \quad i = 1, \ldots, n-1,$$

which gives us $2(n-1)$ equations. Next, we want each cubic function to join as smoothly with its neighbors as possible, so we constrain the splines to have continuous first and second derivatives at the data points $i = 2, \ldots, n-1$.

$$S_i'(x_{i+1}) = S_{i+1}'(x_{i+1}), \quad i = 1, \ldots, n-2,$$

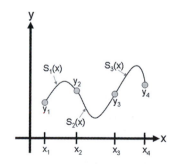

FIGURE 14.3

Illustration of cubic spline interpolation.

$$S_i''(x_{i+1}) = S_{i+1}''(x_{i+1}), \quad i = 1, \ldots, n-2,$$

which gives us $2(n-2)$ equations.

Two more equations are required to compute the coefficients of $S_i(x)$. These last two constraints are arbitrary, and they can be chosen to fit the circumstances of the interpolation being performed. A common set of final constraints is to assume that the second derivatives are zero at the endpoints. This means that the curve is a "straight line" at the end points. Explicitly,

$$S_1''(x_1) = 0$$
$$S_{n-1}''(x_n) = 0.$$

Note that these constraints are not the same as the ones used by MATLAB's interp1 for performing cubic splines, which adds constraints that preserve monotonicity and local extreme points (see the help for interp1 to learn more about this).

TRY IT! Use interp1 to plot the cubic spline interpolation of the data set x = [0 1 2] and y = [1 3 2] for $0 \le x \le 2$. (See Figure 14.4.)

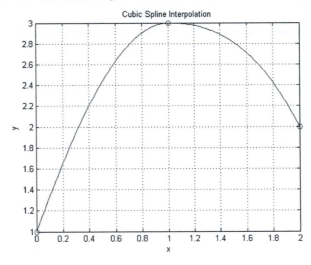

FIGURE 14.4

Resulting plot of previous code. Cubic spline interpolation of the points $x = (0, 1, 2)$ and $y = (1, 3, 2)$.

```
>> X = [0, 1, 2];
>> Y = [1, 3, 2];
>> x = linspace(0,2,100);
>> y = interp1(X,Y,x,'cubic');
>> hold on
```

```
>> plot(x,y)
>> plot(X,Y,'ro')
>> grid on
>> title('Cubic Spline Interpolation')
>> xlabel('x')
>> ylabel('y')
```

To determine the coefficients of each cubic function, we write out the constraints explicitly as a system of linear equations with $4(n-1)$ unknowns. For n data points, the unknowns are the coefficients a_i, b_i, c_i, d_i of the cubic spline, S_i joining the points x_i and x_{i+1}.

For the constraints $S_i(x_i) = y_i$ we have:

$$
\begin{aligned}
a_1 x_1^3 + & b_1 x_1^2 + & c_1 x_1 + & d_1 = & y_1, \\
a_2 x_2^3 + & b_2 x_2^2 + & c_2 x_2 + & d_2 = & y_2,
\end{aligned}
$$
$$\cdots$$
$$a_{n-1} x_{n-1}^3 + b_{n-1} x_{n-1}^2 + c_{n-1} x_{n-1} + d_{n-1} = y_{n-1}.$$

For the constraints $S_i(x_{i+1}) = y_{i+1}$ we have:

$$
\begin{aligned}
a_1 x_2^3 + & b_1 x_2^2 + & c_1 x_2 + & d_1 = y_2, \\
a_2 x_3^3 + & b_2 x_3^2 + & c_2 x_3 + & d_2 = y_3,
\end{aligned}
$$
$$\cdots$$
$$a_{n-1} x_n^3 + b_{n-1} x_n^2 + c_{n-1} x_n + d_{n-1} = y_n.$$

For the constraints $S_i'(x_{i+1}) = S_{i+1}'(x_{i+1})$ we have:

$$
\begin{aligned}
3a_1 x_2^2 + & 2b_1 x_2 + & c_1 - & 3a_2 x_2^2 - & 2b_2 x_2 - & c_2 = 0, \\
3a_2 x_3^2 + & 2b_2 x_3 + & c_2 - & 3a_3 x_3^2 - & 2b_3 x_3 - & c_3 = 0,
\end{aligned}
$$
$$\cdots$$
$$3a_{n-2} x_{n-1}^2 + 2b_{n-2} x_{n-1} + c_{n-2} - 3a_{n-1} x_{n-1}^2 - 2b_{n-1} x_{n-1} - c_{n-1} = 0.$$

For the constraints $S_i''(x_{i+1}) = S_{i+1}''(x_{i+1})$ we have:

$$
\begin{aligned}
6a_1 x_2 + & 2b_1 - & 6a_2 x_2 - & 2b_2 = 0, \\
6a_2 x_3 + & 2b_2 - & 6a_3 x_3 - & 2b_3 = 0, \\
+ & & \cdots -
\end{aligned}
$$
$$6a_{n-2} x_{n-1} + 2b_{n-2} - 6a_{n-1} x_{n-1} - 2b_{n-1} = 0.$$

Finally for the endpoint constraints $S_1''(x_1) = 0$ and $S_{n-1}''(x_n) = 0$, we have:

$$
\begin{aligned}
6a_1 x_1 + & 2b_1 = 0, \\
6a_{n-1} x_n + & 2b_{n-1} = 0.
\end{aligned}
$$

These equations are linear in the unknown coefficients a_i, b_i, c_i, and d_i. We can put them in matrix form and solve for the coefficients of each spline by left division. Remember that whenever we solve the matrix equation $Ax = b$ for x, we must make be sure that A is square and invertible. In the case of finding cubic spline equations, the A matrix is always square and invertible as long as the x_i values in the data set are unique.

TRY IT! Find the cubic spline interpolation at x = 1.5 based on the data x = [0 1 2], y = [1 3 2].

First we create the appropriate system of equations and find the coefficients of the cubic splines by solving the system in matrix form.

The matrix form of the system of equations is:

$$\begin{bmatrix} 0 & 0 & 0 & 1 & 0 & 0 & 0 & 0 \\ 0 & 0 & 0 & 0 & 1 & 1 & 1 & 1 \\ 1 & 1 & 1 & 1 & 0 & 0 & 0 & 0 \\ 0 & 0 & 0 & 0 & 8 & 4 & 2 & 1 \\ 3 & 2 & 1 & 0 & -3 & -2 & -1 & 0 \\ 6 & 2 & 0 & 0 & -6 & -2 & 0 & 0 \\ 0 & 2 & 0 & 0 & 0 & 0 & 0 & 0 \\ 0 & 0 & 0 & 0 & 12 & 2 & 0 & 0 \end{bmatrix} \begin{bmatrix} a_1 \\ b_1 \\ c_1 \\ d_1 \\ a_2 \\ b_2 \\ c_2 \\ d_2 \end{bmatrix} = \begin{bmatrix} 1 \\ 3 \\ 3 \\ 2 \\ 0 \\ 0 \\ 0 \\ 0 \end{bmatrix}$$

By left division, we get the following results from MATLAB:

```
>> x = A\b
x =
   -0.7500
        0
    2.7500
    1.0000
    0.7500
   -4.5000
    7.2500
   -0.5000
```

Therefore, the two cubic polynomials are

$$S_1(x) = -.75x^3 + 2.75x + 1, \quad \text{for } 0 \le x \le 1 \text{ and}$$
$$S_2(x) = .75x^3 - 4.5x^2 + 7.25x - .5, \quad \text{for } 1 \le x \le 2$$

So for $x = 1.5$ we evaluate $S_2(1.5)$ and get an estimated value of 2.7813.

14.4 Lagrange Polynomial Interpolation

Rather than finding cubic polynomials between subsequent pairs of data points, **Lagrange polynomial interpolation** finds a single polynomial that goes through all the data points. This polynomial is referred to as a **Lagrange polynomial**, $L(x)$, and as an interpolation function, it should have the property

$L(x_i) = y_i$ for every point in the data set. For computing Lagrange polynomials, it is useful to write them as a linear combination of **Lagrange basis polynomials**, $P_i(x)$, where

$$P_i(x) = \prod_{j=1, j \neq i}^{n} \frac{x - x_j}{x_i - x_j},$$

and

$$L(x) = \sum_{i=1}^{n} y_i P_i(x).$$

Here, $\prod$ means "the product of" or "multiply out."

You will notice that by construction, $P_i(x)$ has the property that $P_i(x_j) = 1$ when $i = j$ and $P_i(x_j) = 0$ when $i \neq j$. Since $L(x)$ is a sum of these polynomials, you can observe that $L(x_i) = y_i$ for every point, exactly as desired.

TRY IT! Find the Lagrange basis polynomials for the data set x = [0 1 2] and y = [1 3 2]. Plot each polynomial (Figure 14.5) and verify the property that $P_i(x_j) = 1$ when $i = j$ and $P_i(x_j) = 0$ when $i \neq j$.

$$P_1(x) = \frac{(x - x_2)(x - x_3)}{(x_1 - x_2)(x_1 - x_3)} = \frac{(x - 1)(x - 2)}{(0 - 1)(0 - 2)} = \frac{1}{2}(x^2 - 3x + 2),$$

$$P_2(x) = \frac{(x - x_1)(x - x_3)}{(x_2 - x_1)(x_2 - x_3)} = \frac{(x - 0)(x - 2)}{(1 - 0)(1 - 2)} = -x^2 + 2x,$$

$$P_3(x) = \frac{(x - x_1)(x - x_2)}{(x_3 - x_1)(x_3 - x_2)} = \frac{(x - 0)(x - 1)}{(2 - 0)(2 - 1)} = \frac{1}{2}(x^2 - x).$$

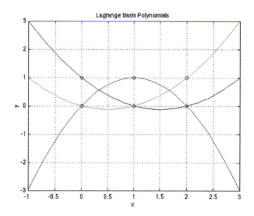

FIGURE 14.5

Lagrange basis polynomials for test data. By design, $P_i(x_j) = 1$ when $i = j$, and $P_i(x_j) = 0$ when $i \neq j$.

```
>> X = [0 1 2];
>> Y = [1 3 2];

>> P1 = @(x) .5*(x.^2 - 3*x + 2);
>> P2 = @(x) -x.^2 + 2*x;
>> P3 = @(x) .5*(x.^2 - x);

>> x = -1:0.1:3;
>> plot(x, P1(x), 'b', x, P2(x), 'r', x, P3(x), 'g')
>> hold on
>> plot(X, ones(size(X)), 'ko', X, zeros(size(X)), 'ko')
>> grid on
>> title('Lagrange Basis Polynomials')
>> xlabel('x')
>> ylabel('y')
```

TRY IT! For the previous example, compute and plot the Lagrange polynomial (Figure 14.6) and verify that it goes through each of the data points.

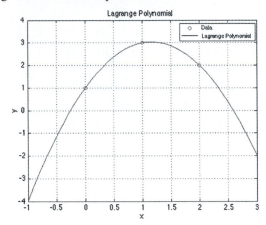

FIGURE 14.6

Resulting plot of previous code. As expected, the Lagrange polynomials goes through each of the data points.

```
>> L = @(x) P1(x) + 3*P2(x) + 2*P3(x);

>> plot(X,Y, 'ro', x, L(x), 'b')
>> grid on
>> title('Lagrange Polynomial')
>> xlabel('x')
>> ylabel('y')
>> legend('Data', 'Lagrange Polynomial')
```

> **WARNING!** Lagrange interpolation polynomials are defined outside the area of interpolation— outside of the interval $[x_1, x_n]$—but grow very fast and unbounded outside this region. This is not a desirable feature because in general, this is not the behavior of the underlying data. Thus, a Lagrange interpolation should never be used to interpolate outside this region.

Summary

1. Given a set of reliable data points, interpolation is a method of estimating dependent variable values for independent variable values not in our data set.
2. Linear, Cubic Spline, and Lagrange interpolation are common interpolating methods.

Vocabulary

cubic-spline interpolation   Lagrange basis polynomial   Lagrange polynomial interpolation
interpolate   Lagrange polynomial   linearly interpolation
interpolation function

Functions and Operators

`interp1`

Problems

1. Write a function with header `[Y] = myLinInterp(x,y,X)`, where x and y are column vectors containing experimental data points, and X is an array. Assume that x and X are in ascending order and have unique elements. The output argument, Y, should be a vector, the same size as X, where `Y(i)` is the linear interpolation of `X(i)`. You may not use `interp1`.

2. Write a function with header `[Y] = myCubicSpline(x,y,X)`, where x and y are column vectors containing experimental data points, and X is an array. Assume that x and X are in ascending order and have unique elements. The output argument, Y, should be a vector, the same size as X, where `Y(i)` is the cubic spline interpolation of `X(i)`. You may not use `interp1`.

3. Write a function with header `[Y] = myNearestNeighbor(x,y,X)`, where x and y are column vectors containing experimental data points, and X is an array. Assume that x and X are in ascending order and have unique elements. The output argument, Y, should be a vector, the same size as X, where `Y(i)` is the nearest neighbor interpolation of `X(i)`. That is, `Y(i)` should be the `y(j)` where `x(j)` is the closest independent data point to `X(i)`. You may not use `interp1`.

4. Think of a situation where using nearest neighbor interpolation would be superior to cubic spline interpolation.

⌊m⌋ **5.** Write a function with header $[Y] = \text{myCubicSplineFlat}(x,y,X)$, where x and y are column vectors containing experimental data points, and X is an array. Assume that x and X are in ascending order and have unique elements. The output argument, Y, should be a vector, the same size as X, where Y(i) is the cubic spline interpolation of X(i). *However*, instead of the standard "clamped" endpoint constraints, use $S_1'(x_1) = 0$ and $S_{n-1}'(x_n) = 0$.

⌊m⌋ **6.** Write a function with header $[Y] = \text{myQuinticSpline}(x,y,X)$, where x and y are column vectors containing experimental data points, and X is an array. Assume that x and X are in ascending order and have unique elements. The output argument, Y, should be a vector, the same size as X, where Y(i) is the quintic spline interpolation of X(i). You will need to use additional endpoint constraints to come up with enough constraints. You may use endpoint constraints at your discretion.

⌊m⌋ **7.** Write a function with header $[] = \text{myInterpPlotter}(x,y,\ X,\ \text{option})$, where x and y are column vectors containing x and y data points, and X is a column vector containing the coordinates for which an interpolation is desired. The input argument option should be a string, either 'linear', 'spline', or 'nearest'. Your function should produce a plot of the data points (x, y) marked as red circles, and the points (X, Y), where X is the input vector and Y is the interpolation at the points contained in X defined by the input argument specified by option. The points (X, Y) should be connected by a blue line. Be sure to include title, axis labels, *and* a legend. Hint: You should use the function interp1.

Test Cases:

```
>> x = [0 .1 .15 .35 .6 .7 .95 1];
>> y = [1 0.8187 0.7408 0.4966 0.3012 0.2466 0.1496 0.1353];
>> myInterpPlotter(x,y, linspace(0,1,100), 'nearest')
```

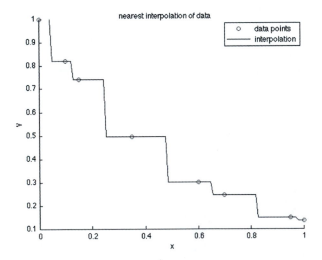

```
>> myInterpPlotter(x,y, linspace(0,1,100), 'linear')
```

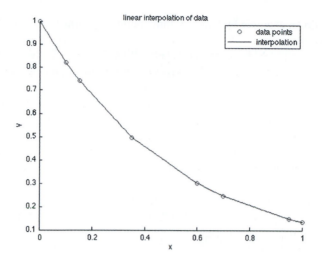

```
>> myInterpPlotter(x,y, linspace(0,1,100), 'cubic')
```

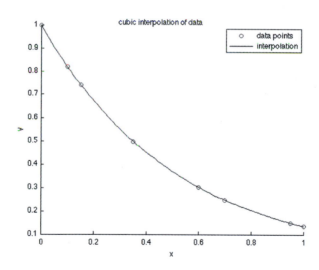

.m **8.** Write a function with header [Y] = myDCubicSpline(x,y, X, D), where Y is the cubic spline interpolation at X taken from the data points contained in x and y. However, instead of the standard pinned endpoint conditions (i.e., $S_1''(x_1) = 0$ and $S_{n-1}''(x_n) = 0$) you should use the endpoint conditions $S_1'(x_1) = D$ and $S_{n-1}'(x_n) = D$ (i.e., the slopes of the interpolating polynomials at the endpoints is D). Note that there may be more than five data points. It may be a good idea to do an example by hand for three points (2 splines).

$$S_i'(x_{i+1}) = S_{i+1}'(x_{i+1}) \quad \text{for } i = 1, \ldots, n-2$$
$$S_i''(x_{i+1}) = S_{i+1}''(x_{i+1}) \quad \text{for } i = 1, \ldots, n-2.$$

Test Cases:

```
>> x = [0 1 2 3 4];
>> y = [0 0 1 0 0];
>> X = linspace(0,4,100);
>> Y = myDCubicSpline(x, y, 1.5, 1)
Y =
      0.5402
```

```
>> subplot(2,2,1)
>> plot(x,y,'ro', X, myDCubicSpline(x,y,X,0))
>> subplot(2,2,2)
>> plot(x,y,'ro', X, myDCubicSpline(x,y,X,1))
>> subplot(2,2,3)
>> plot(x,y,'ro', X, myDCubicSpline(x,y,X,-1))
>> subplot(2,2,4)
>> plot(x,y,'ro', X, myDCubicSpline(x,y,X,4))
```

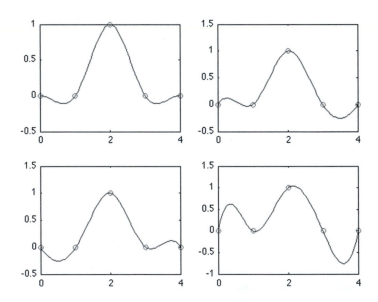

9. Write a function with header [Y] = myLagrange(x,y, X), where Y is the Lagrange interpolation of the data points contained in x and y computed at X. Hint: Use a nested for-loop, where the inner for-loop computes the product for the Lagrange basis polynomial and the outer loop computes the sum for the Lagrange polynomial.

Test Cases:

```
>> x = [0 1 2 3 4];
>> y = [2 1 3 5 1];
>> Y = myLagrange(x,y, 1.5)
Y =
    0.7031
>> X = linspace(0,4,100);
>> hold on
>> plot(x,y,'ro')
>> plot(X, myLagrange(x,y, X))
>> xlabel('x')
>> ylabel('y')
>> title('Lagrange Interpolation of Data Points')
```

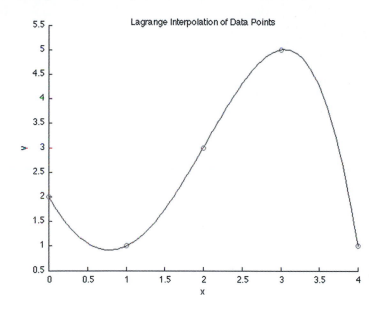

Series

15

CHAPTER OUTLINE

Motivation

Many functions such as $\sin(x)$ and $\cos(x)$ are useful for engineers, but they are impossible to compute explicitly. In practice, these functions can be approximated by sums of functions that are easy to compute, such as polynomials. In fact, most functions common to engineers cannot be computed without approximations of this kind. Since these functions are used so often, it is important to know how these approximations work and their limitations.

In this chapter, you will learn about Taylor series, which is one method of approximating complicated functions. This chapter in no way takes the place of a full course on functional analysis, but it does provide exposure that will be useful for subsequent chapters.

15.1 Expressing Functions with Taylor Series

A **sequence** is an ordered set of numbers denoted by the list of numbers inside parentheses. For example, $s = (s_1, s_2, s_3, \cdots)$ means s is the sequence $s_1, s_2, s_3, \cdots$ and so on. In this context, "ordered" means that s_1 comes *before* s_2, not that $s_1 < s_2$. Many sequences have a more complicated structure. For example, $s = (n^2, n \in \mathbb{N})$ is the sequence $0, 1, 4, 9, \cdots$. A **series** is the sum of a sequence up to a certain element. An **infinite sequence** is a sequence with an infinite number of terms, and an **infinite series** is the sum of an infinite sequence.

A **Taylor series expansion** is a representation of a function by an infinite series of polynomials around a point. Mathematically, the Taylor series of a function, $f(x)$, is defined as:

$$f(x) = \sum_{n=0}^{\infty} \frac{f^{(n)}(a)(x-a)^n}{n!},$$

where $f^{(n)}$ is the nth derivative of f and $f^{(0)}$ is the function f.

An Introduction to MATLAB® Programming and Numerical Methods. http://dx.doi.org/10.1016/B978-0-12-420228-3.00015-4

TRY IT! Compute the Taylor series expansion for $f(x) = 5x^2 + 3x + 5$ around $a = 0$, and $a = 1$. Verify that f and its Taylor series expansions are identical.

First compute derivatives analytically:

$$f(x) = 5x^2 + 3x + 5$$
$$f'(x) = 10x + 3$$
$$f''(x) = 10$$

Around a = 0:

$$f(x) = \frac{5x^0}{0!} + \frac{3x^1}{1!} + \frac{10x^2}{2!} + 0 + 0 + \cdots = 5x^2 + 3x + 5$$

Around a = 1:

$$f(x) = \frac{13(x-1)^0}{0!} + \frac{13(x-1)^1}{1!} + \frac{10(x-1)^2}{2!} + 0 + \cdots$$
$$= 13 + 13x - 13 + 5x^2 - 10x + 5 = 5x^2 + 3x + 5$$

Note: The Taylor series expansion of any polynomial has finite terms because the nth derivative of any polynomial is 0 for n large enough.

TRY IT! Write the Taylor series for $\sin(x)$ around the point $a = 0$.

Let $f(x) = \sin(x)$. Then according to the Taylor series expansion,

$$f(x) = \frac{\sin(0)}{0!}x^0 + \frac{\cos(0)}{1!}x^1 + \frac{-\sin(0)}{2!}x^2 + \frac{-\cos(0)}{3!}x^3 + \frac{\sin(0)}{4!}x^4 + \frac{\cos(0)}{5!}x^5 + \cdots.$$

The expansion can be written compactly by the formula

$$f(x) = \sum_{n=0}^{\infty} \frac{(-1)^n x^{2n+1}}{(2n+1)!},$$

which ignores the terms that contain $\sin(0)$ (i.e., the even terms). However, because these terms are ignored, the terms in this series and the proper Taylor series expansion are off by a factor of $2n + 1$; for example the $n = 0$ term in formula is the $n = 1$ term in the Taylor series, and the $n = 1$ term in the formula is the $n = 3$ term in the Taylor series.

15.2 Approximations with Taylor Series

Clearly, it is not useful to express functions as infinite sums because we cannot even compute them that way. However, it is often useful to approximate functions by using an Nth **order Taylor series**

approximation of a function, which is a truncation of its Taylor expansion at some $n = N$. This technique is especially powerful especially when there is a point around which we have knowledge about a function for all its derivatives. For example, if we take the Taylor expansion of e^x around $a = 0$, then $f^{(n)}(a) = 1$ for all n, we don't even have to compute the derivatives in the Taylor expansion to approximate e^x!

TRY IT! Use MATLAB to plot the sin function along with the first, third, fifth, and seventh order Taylor series approximations. Note that this is the zero-th to third term in the formula given earlier (See Figure 15.1.).

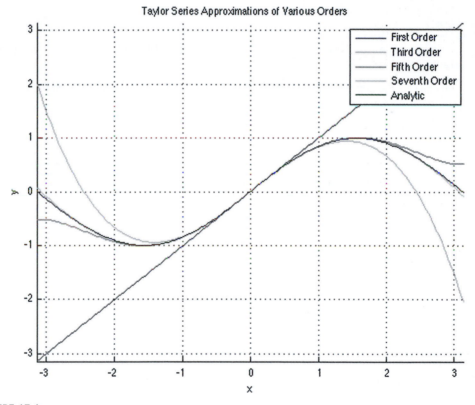

FIGURE 15.1

Resulting plot of previous code. Successive orders of approximation of the sin function by its Taylor expansion.

```
clc
clear
close all
```

```
c = {'b','g','r','c'};
x = linspace(-pi, pi, 200);
y = zeros(size(x));

hold on
for n = 0:3
    y = y + ((-1)^n * x.^(2*n + 1))/(factorial(2*n + 1));
    plot(x,y,c{n+1})
end

plot(x,sin(x),'k')
grid on
title('Taylor Series Approximations of Various Orders')
xlabel('x')
ylabel('y')
legend('First Order', 'Third Order', 'Fifth Order', 'Seventh Order', 'Analytic')
axis tight
```

As you can see, the approximation approaches the analytic function quickly, even for x not near to $a = 0$.

TRY IT! Compute the seventh order Taylor series approximation for $\sin(x)$ around $a = 0$ at $x = \pi/2$. Compare the value to the correct value, 1.

```
>> x = pi/2;
>> y = 0;
>> for i = 0:3
        y = y + ((-1)^i * x^(2*i + 1))/(factorial(2*i + 1));
    end
>> format long
>> y
y =
    0.999843101399499
```

The seventh order Taylor series approximation is very close to the theoretical value of the function even if it is computed far from the point around which the Taylor series was computed (i.e., $x = \pi/2$ and $a = 0$).

The most common Taylor series approximation is the first order approximation, or **linear approximation**. Intuitively, for "smooth" functions the linear approximation of the function around a point, a, can be made as good as you want provided you stay sufficiently close to a. In other words, "smooth" functions look more and more like a line the more you zoom into any point. This fact is depicted in Figure 15.2. Linear approximations are useful tools when analyzing complicated functions locally.

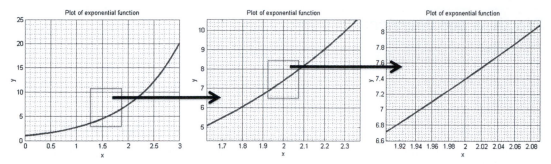

FIGURE 15.2

Successive levels of zoom of a smooth function to illustrate the linear nature of functions locally.

TRY IT! Take the linear approximation for e^x around the point $a = 0$. Use the linear approximation for e^x to approximate the value of e^1 and $e^{0.01}$. Use MATLAB's function exp to compute exp(1) and exp(0.01) for comparison.

The linear approximation of e^x around $a = 0$ is $1 + x$.

MATLAB's exp function gives the following:

```
>> exp(1)
ans =
    2.718281828459046
>> exp(0.01)
ans =
    1.010050167084168
```

The linear approximation of e^1 is 2, which is inaccurate, and the linear approximation of $e^{0.01}$ is 1.01, which is very good. This example illustrates how the linear approximation becomes close to the functions close to the point around which the approximation is taken.

Summary

1. Some functions can be perfectly represented by a Taylor series, an infinite sum of polynomials.
2. Functions that have a Taylor series expansion can be approximated by truncating its Taylor series.
3. The linear approximation is a common local approximation for functions.

Vocabulary

infinite sequence N^{th} order Taylor series approximation   Taylor series expansion
infinite series   sequence
linear approximation   series

Functions and Operators

(none)

Problems

1. Use Taylor series expansions to show that $e^{ix} = \cos(x) + i\sin(x)$, where $i = \sqrt{-1}$.

2. Use the linear approximation of $\sin(x)$ around $a = 0$ to show that $\frac{\sin(x)}{x} \approx 1$ for small x.

3. Write the Taylor series expansion for e^{x^2} around $a = 0$. Write a function with header $[\text{approx}] = \text{myDoubleExp}(x, N)$, which computes an approximation of e^{x^2} using the first N terms of the Taylor series expansion. Be sure that myDoubleExp can take array inputs.

4. Write a function that gives the Taylor series approximation to the exp function around 0 for order 1 through 7.

5. Compute the fourth order Taylor expansion for $\sin(x)$ and $\cos(x)$ and $\sin(x)\cos(x)$ around 0. Which produces less error for $x = \pi/2$: computing the Taylor expansion for sin and cos separately then multiplying the result together, or computing the Taylor expansion for the product first then plugging in x?

6. Write a function with header $[\text{yApprox}] = \text{myCoshApproximator}(x, n)$, where yApprox is the n-th order Taylor Series approximation for $\cosh(x)$, the hyperbolic cosine of x, taken around $a = 0$. You may assume that x is a vector and n is a positive integer (including 0). Note that your function should be vectorized for x. Recall that

$$\cosh(x) = (e^x + e^{-x})/2.$$

Warning: The approximations for $n = 0$ and $n = 1$ will be equivalent, the approximations for $n = 2$ and $n = 3$ will be equivalent, and so on.

Test Cases:

```
>> format long
>> yTrue = cosh(1.5)
yTrue =
     2.352409615243247
>> y3 = myCoshApproximator(1.5, 3)
y3 =
     2.125000000000000
>> y10 = myCoshApproximator(1.5, 10)
y10 =
     2.352409340994699
>> y20 = myCoshApproximator(1.5, 20)
y20 =
     2.352409615243247
>> x = linspace(0,2,100);
```

```
>> plot(x, cosh(x), x, myCoshApproximator(x,10))
>> xlabel('x')
>> ylabel('y')
>> title('cosh(x) with 10th order approximation')
>> plot(x, cosh(x), x, myCoshApproximator(x,2))
>> title('cosh(x) with 2nd order approximation')
>> legend('real', 'approx')
```

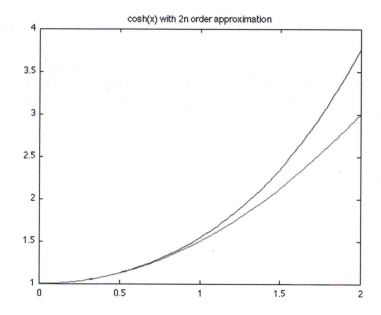

Root Finding

16

CHAPTER OUTLINE

Motivation

As the name suggests, the roots of a function are one of its most important properties. Finding the roots of functions is important in many engineering applications such as signal processing and optimization. For simple functions such as $f(x) = ax^2 + bx + c$, you may already be familiar with the "quadratic formula,"

$$x_r = \frac{-b \pm \sqrt{b^2 - 4ac}}{2a},$$

which gives x_r, the two roots of f exactly. However for more complicated functions, the roots can rarely be computed using such explicit, or exact, means.

By the end of this chapter, you should understand the root finding problem, and two algorithms for finding roots to functions, their properties, and their limitations.

16.1 Root Finding Problem Statement

The **root** or **zero** of a function, $f(x)$, is an x_r such that $f(x_r) = 0$. For functions such as $f(x) = x^2 - 9$, the roots are clearly 3 and -3. However, for other functions such as $f(x) = \cos(x) - x$, determining an **analytic**, or exact, solution for the roots of functions can be difficult. For these cases, it is useful to generate numerical approximations of the roots of f and understand the limitations in doing so.

An Introduction to MATLAB® Programming and Numerical Methods. http://dx.doi.org/10.1016/B978-0-12-420228-3.00016-6

TRY IT! Use the `fzero` function to compute the root of $f(x) = \cos(x) - x$ near -2. Verify that the solution given by `fzero` is a root (or close enough).

```
>> f = @(x) cos(x) - x;
>> r = fzero(f,-2)
r =
     0.7391
>> F = f(r)
F =
    -2.4881e-005
```

TRY IT! The function $f(x) = \frac{1}{x}$ has no root. Use the `fzero` function to try to compute the root of $f(x) = \frac{1}{x}$. Verify that the solution given from `fzero` is not a root of f by plugging the estimated root value back into f.

```
>> f = @(x) 1./x;
>> r = fzero(f,-2)
r =
    -2.6773e-16
>> F = f(r)
F =
    -3.7351e+015
```

Clearly F is large (i.e., $f(x) \neq 0$) and so r is not a root. Understanding why a MATLAB function would make this mistake requires knowledge about the function being analyzed and understanding how the algorithms for finding roots work.

16.2 Tolerance

In engineering, **error** is a deviation from an expected or computed value. **Tolerance** is the level of error that is acceptable for an engineering application. We say that a computer program has **converged** to a solution when it has found a solution with an error smaller than the tolerance. When computing roots numerically, or conducting any other kind of numerical analysis, it is important to establish both a metric for error and a tolerance that is suitable for a given engineering application.

For computing roots, we want an x_r such that $f(x_r)$ is very close to 0. Therefore $|f(x)|$ is a possible choice for the measure of error since the smaller it is, the likelier we are to a root. Also if we assume that x_i is the ith guess of an algorithm for finding a root, then $|x_{i+1} - x_i|$ is another possible choice for measuring error, since we expect the improvements between subsequent guesses to diminish as it approaches a solution. As will be demonstrated in the following examples, these different choices have their advantages and disadvantages.

EXAMPLE: Let error be measured by $e = |f(x)|$ and tol be the acceptable level of error. The function $f(x) = x^2 + \text{tol}/2$ has no real roots. However, $|f(0)| = \text{tol}/2$ and is therefore acceptable as a solution for a root finding program.

EXAMPLE: Let error be measured by $e = |x_{i+1} - x_i|$ and tol be the acceptable level of error. The function $f(x) = 1/x$ has no real roots, but the guesses $x_i = -\text{tol}/4$ and $x_{i+1} = \text{tol}/4$ have an error of $e = \text{tol}/2$ and is an acceptable solution for a computer program.

Based on these observations, the use of tolerance and converging criteria must be done very carefully and in the context of the program that uses them.

16.3 Bisection Method

The **Intermediate Value Theorem** says that if $f(x)$ is a continuous function between a and b, and $\text{sign}(f(a)) \neq \text{sign}(f(b))$, then there must be a c, such that $a < c < b$ and $f(c) = 0$. This is illustrated in Figure 16.1.

The **bisection method** uses the intermediate value theorem iteratively to find roots. Let $f(x)$ be a continuous function, and a and b be real scalar values such that $a < b$. Assume, without loss of generality, that $f(a) > 0$ and $f(b) < 0$. Then by the intermediate value theorem, there must be a root on the open interval (a, b). Now let $m = \frac{b+a}{2}$, the midpoint between and a and b. If $f(m) = 0$ or is close enough, then m is a root. If $f(m) > 0$, then m is an improvement on the left bound, a, and there is guaranteed to be a root on the open interval (m, b). If $f(m) < 0$, then m is an improvement on the right bound, b, and there is guaranteed to be a root on the open interval (a, m). This scenario is depicted in Figure 16.2.

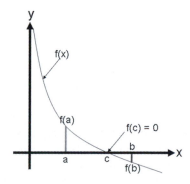

FIGURE 16.1

Illustration of intermediate value theorem. If $\text{sign}(f(a)) \neq \text{sign}(f(b))$, then $\exists c \in (a, b)$ such that $f(c) = 0$.

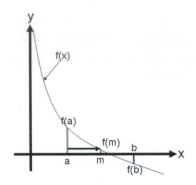

FIGURE 16.2

Illustration of the bisection method. The sign of $f(m)$ is checked to determine if the root is contained in the interval (a, m) or (m, b). This new interval is used in the next iteration of the bisection method. In the case depicted in the figure, the root is in the interval (m,b).

The process of updating a and b can be repeated until the error is acceptably low.

TRY IT! Program a function with header [R] = myBisection(f,a,b,tol) that approximates a root R of f, bounded by a and b to within $|f(\frac{a+b}{2})| <$ tol.

```
function [R] = myBisection(f, a, b, tol)
% [R] = myBisection(f, a, b, tol)
% Approximates a root, R, of f bounded by a and b to within tolerance
% |f(m)| < tol with m the midpoint between a and b.
% Recursive implementation (for fun)

% check if a and b bound a root
if sign(f(a)) == sign(f(b))
    error('The scalars a and b do not bound a root')
end

% get midpoint
m = (a + b)/2;

if abs(f(m)) < tol
    % stopping condition, report m as root
    R = m;
elseif sign(f(a)) == sign(f(m))
    % case where m is improvement on a. Make recursive call with a = m.
    R = myBisection(f, m, b, tol);
elseif sign(f(b)) == sign(f(m))
    % case where m is improvement on b. Make recursive call with b = m.
    R = myBisection(f, a, m, tol);
end

end % end myBisection
```

TRY IT! The $\sqrt{2}$ can be computed as the root of the function $f(x) = x^2 - 2$. Starting at $a = 0$ and $b = 2$, use `myBisection` to approximate the $\sqrt{2}$ to a tolerance of $|f(x)| < 0.1$ and $|f(x)| < 0.01$. Verify that the results are close to a root by plugging the root back into the function.

```
>> format long
>> R = sqrt(2)
R =
   1.414213562373095
>> f = @(x) x^2 - 2;
>> r1 = myBisection(f, 0, 2, .1)
r1 =
   1.437500000000000
>> r01 = myBisection(f, 0, 2, .01)
r01 =
   1.414062500000000
>> f(r1)
ans =
   0.066406250000000
EDU>> f(r01)
ans =
  -4.272460937500000e-004
```

16.4 Newton-Raphson Method

Let $f(x)$ be a smooth and continuous function and x_r be an unknown root of $f(x)$. Now assume that x_0 is a guess for x_r. Unless x_0 is a very lucky guess, $f(x_0)$ will not be a root. Given this scenario, we want to find an x_1 that is an improvement on x_0 (i.e., closer to x_r than x_0). If we assume that x_0 is "close enough" to x_r, then we can improve upon it by taking the linear approximation of $f(x)$ around x_0, which is a line, and finding the intersection of this line with the x-axis. Written out, the linear approximation of $f(x)$ around x_0 is $f(x) \approx f(x_0) + f'(x_0)(x - x_0)$. Using this approximation, we find x_1 such that $f(x_1) = 0$. Plugging these values into the linear approximation results in the equation

$$0 = f(x_0) + f'(x_0)(x_1 - x_0),$$

which when solved for x_1 is

$$x_1 = x_0 - \frac{f(x_0)}{f'(x_0)}.$$

An illustration of how this linear approximation improves an initial guess is shown in Figure 16.3.

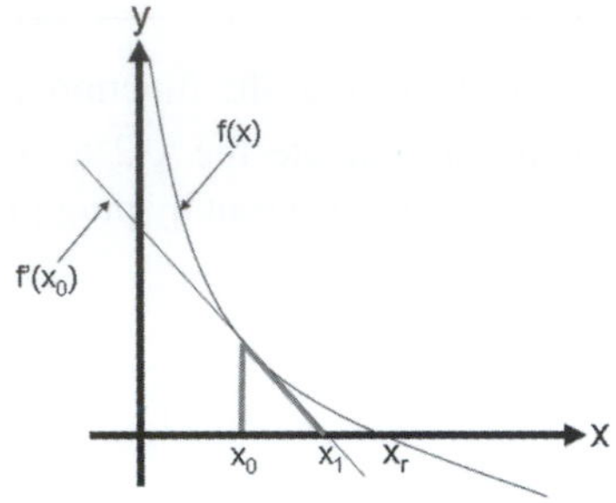

FIGURE 16.3

Illustration of Newton step for a smooth function, $g(x)$.

Written generally, a **Newton step** computes an improved guess, x_i, using a previous guess x_{i-1}, and is given by the equation

$$x_i = x_{i-1} - \frac{g(x_{i-1})}{g'(x_{i-1})}.$$

The **Newton-Raphson Method** of finding roots iterates Newton steps from x_0 until the error is less than the tolerance.

TRY IT! Again, the $\sqrt{2}$ is the root of the function $f(x) = x^2 - 2$. Using $x_0 = 1.4$ as a starting point, use the previous equation to estimate $\sqrt{2}$. Compare this approximation with the value computed by MATLAB's sqrt function.

$$x = 1.4 - \frac{1.4^2 - 2}{2(1.4)} = 1.414285714285714$$

```
>> format long
>> sqrt(2)
ans =
      1.414213562373095
```

TRY IT! Write a function with header [R] = myNewton(f, df, x0, tol), where R is an estimation of the root of f, f is a handle to the function $f(x)$, df is a handle to the function $f'(x)$, x0 is an initial guess, and tol is the error tolerance. The error measurement should be $|f(x)|$.

```
function [R] = myNewton(f, df, x0, tol)
% [R] = myNewton(f, df, x0, tol)
% R is an estimation of the root of f using the Newton Raphson method.
% Recursive implementation (for fun).
%

if abs(f(x0)) < tol
    R = x0;
else
    R = myNewton(f, df, x0 - f(x0)/df(x0), tol);
end

end % end myNewton
```

TRY IT! Use myNewton to compute $\sqrt{2}$ to within tolerance of 1e-6 starting at x0 = 1.5.

```
>> format long
>> R = myNewton(@(x) x.^2 - 2, @(x) 2*x, 1.5, 1e-6)
R =
    1.414213562374690
>> sqrt(2)
ans =
    1.414213562373095
```

If x_0 is close to x_r, then it can be proven that, in general, the Newton-Raphson method converges to x_r much faster than the bisection method. However since x_r is initially unknown, there is no way to know if the initial guess is close enough to the root to get this behavior unless some special information about the function is known *a priori* (e.g., the function has a root close to $x = 0$). In addition to this initialization problem, the Newton-Raphson method has other serious limitations. For example, if the derivative at a guess is close to 0, then the Newton step will be very large and probably lead far away from the root. Also, depending on the behavior of the function derivative between x_0 and x_r, the Newton-Raphson method may converge to a different root than x_r that may not be useful for our engineering application.

TRY IT! Compute a single Newton step to get an improved approximation of the root of the function $f(x) = x^3 + 3x^2 - 2x - 5$ and an initial guess, $x_0 = 0.29$.

```
>> x0 = .29;
>> x1 = x0 - (x0.^3 + 3*x0.^2 - 2.*x0 - 5)/(3*x0.^2 + 6*x0 - 2)
x1 =
  -688.4517
```

Note that $f'(x_0) = -0.0077$ (close to 0) and the error at x_1 is approximately 324880000 (very large).

> **TRY IT!** Consider the polynomial $f(x) = x^3 - 100x^2 - x + 100$. This polynomial has a root at $x = 1$ and $x = 100$. Use the Newton-Raphson to find a root of f starting at $x_0 = 0$.
>
>   At $x_0 = 0$, $f(x_0) = 100$, and $f'(x) = -1$. A Newton step gives $x_1 = 0 - \frac{100}{-1} = 100$, which is a root of f. However, note that this root is much farther from the initial guess than the other root at $x = 1$, and it may not be the root you wanted from an initial guess of 0.

Summary

1. Roots are an important property of functions.
2. The bisection method is a way of finding roots based on divide and conquer. Although stable, it might converge slowly compared to the Newton-Raphson method.
3. The Newton-Raphson method is a different way of finding roots based on approximation of the function. The Newton-Raphson method converges quickly close to the actual root, but can have unstable behavior.

Vocabulary

analytic	Intermediate Value Theorem	tolerance
bisection method	Newton-Raphson method	zero
converge	Newton step	
error	root	

Functions and Operators

fzero   roots

Problems

⌨ 1. Write a function with header [R] = myNthRoot(x, N, tol), where x and tol are strictly positive scalars, and N is an integer strictly greater than 1. The output argument, R, should be an approximation $R = \sqrt[N]{x}$, the N-th root of x. This approximation should be computed by using the Newton Raphson method to find the root of the function $f(y) = y^N - x$. The error metric should be $|f(y)|$.

⌨ 2. Write a function with header [X] = myFixedPoint(f,g,tol,maxIter), where f and g are function handles and tol and maxIter are strictly positive scalars. The input argument, maxIter, is also an integer. The output argument, X, should be a scalar satisfying $|f(X) - g(X)| < tol$; that is, X is a point that (almost) satisfies f(X) = g(X). To find X, you should use the Bisection method with error metric, $|F(m)| < tol$. The function myFixedPoint should "give up" after maxIter number of iterations and return X = [] if this occurs.

⌊m 3. Why does the bisection method fail for $f(x) = 1/x$ with error given by $|b - a|$? Hint: How does $f(x)$ violate the intermediate value theorem?

⌊m 4. Write a function with header [R, E] = myBisection(f, a, b, tol), where f is a function handle, a and b are scalars such that $a < b$, and tol is a strictly positive scalar value. The function should return an array, R, where $R(i)$ is the estimation of the root of f defined by $(a+b)/2$ for the i-th iteration of the bisection method. Remember to include the initial estimate. The function should also return an array, E, where $E(i)$ is the value of $|f(R(i))|$ for the i-th iteration of the bisection method. The function should terminate when $E(i) < tol$. You may assume that $\text{sign}(f(a)) \neq \text{sign}(f(b))$.

Clarification: The input a and b constitute the first iteration of bisection, and therefore R and E should never be empty.

Test Cases:

```
>> f = @(x) x.^2 - 2;
>> [R, E] = myBisection(f, 0, 2, 1e-1)

R =
      1.0000    1.5000    1.2500    1.3750    1.4375
E =
      1.0000    0.2500    0.4375    0.1094    0.0664


>> f = @(x) sin(x) - cos(x);
>> [R, E] = myBisection(f, 0, 2, 1e-2)
R =
      1.0000    0.5000    0.7500    0.8750    0.8125    0.7813
E =
      0.3012    0.3982    0.0501    0.1265    0.0383    0.0059
```

⌊m 5. Write a function with header [R, E] = myNewton(f, df, x0, tol) where f is a function handle, df is a function handle to the derivative of f, x0 is an initial estimation of the root, and tol is a strictly positive scalar. The function should return an array, R, where $R(i)$ is the Newton-Raphson estimation of the root of f for the i-th iteration. Remember to include the initial estimate. The function should also return an array, E, where $E(i)$ is the value of $|f(R(i))|$ for the i-th iteration of the Newton-Raphson method. The function should terminate when $E(i) < tol$. You may assume that the derivative of f will not hit 0 during any iteration for any of the test cases given.

Test Cases:

```
>> f = @(x) x^2 - 2;
>> df = @(x) 2*x;
>> [R, E] = myNewton(f, df, 1, 1e-5)
R =
     1.000000000000000   1.500000000000000   1.416666666666667   1.414215686274510
E =
     1.000000000000000   0.250000000000000   0.006944444444445   0.000006007304883
```

```
f = @(x) sin(x) − cos(x);
df = @(x) cos(x) + sin(x);
>> [R, E] = myNewton(f, df, 1, 1e−5)
R =
    1.000000000000000    0.782041901539138    0.785398175999702
E =
    0.301168678939757    0.004746462127804    0.000000017822278
```

Test Cases:

```
>> f = @(x) exp(x) − 3;
>> df = @(x) exp(x);
>> g = @(x) sqrt(x);
>> dg = @(x) .5*x^(−.5);
>> x = myNLEQ(f, df, g, dg, 1, 1e−6)
x =
    1.434542442506692

>> p1 = [1 −2 3 −8];
>> p2 = [1 −3 2 −4];
>> f = @(x) polyval(p1,x);
>> df = @(x) polyval(polyder(p1),x);
>> g = @(x) polyval(p2,x);
>> dg = @(x) polyval(polyder(p2),x);
>> x = myNLEQ(f, df, g, dg, 2, 1e−6)
x =
    1.561552842846145
```

.m **6.** Consider the problem of building a pipeline from an offshore oil platform, a distance H miles from the shoreline, to an oil refinery station on land, a distance L miles along the shore. The cost of building the pipe is $C_{ocean/mile}$ while the pipe is under the ocean and $C_{land/mile}$ while the pipe is on land. The pipe will be built in a straight line toward the shore where it will make contact at some point, x, between 0 and L. It will continue along the shore on land until it reaches the oil refinery. See the figure for clarification.

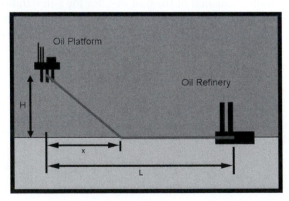

Write a function with header [x] = myPipeBuilder(C_ocean, C_land, L, H), where the input arguments are as described earlier, and x is the x-value that minimizes the total cost of the pipeline. You should use the bisection method to determine this value to within a tolerance of $1 \cdot 10^{-6}$ starting at an initial bound of $a = 0$ and $b = L$.

Test Cases:

```
>> [x] = myPipeBuilder(20, 10, 100, 50)
x =
   28.867512941360474
>> [x] = myPipeBuilder(30, 10, 100, 50)
x =
   17.677670717239380
>> [x] = myPipeBuilder(30, 10, 100, 20)
x =
    7.071067392826080
```

7. Find a function $f(x)$ and guess for the root of f, x_0, such that the Newton-Raphson method would oscillate between x_0 and $-x_0$ indefinitely.

Numerical Differentiation

CHAPTER OUTLINE

Motivation

Many engineering systems change over time, space, and many other dimensions of interest. In mathematics, function derivatives are often used to model these changes. However, in practice the function may not be explicitly known, or the function may be implicitly represented by a set of data points. In these cases and others, it may be desirable to compute derivatives numerically rather than analytically.

The focus of this chapter is numerical differentiation. By the end of this chapter you should be able to derive some basic numerical differentiation schemes and their accuracy.

17.1 Numerical Differentiation Problem Statement

A **numerical grid** is an evenly spaced set of points over the domain of a function (i.e., the independent variable), over some interval. The **spacing** or **step size** of a numerical grid is the distance between adjacent points on the grid. For the purpose of this text, if x is a numerical grid, then x_j is the jth point in the numerical grid and h is the spacing between x_{j-1} and x_j. Figure 17.1 shows an example of a numerical grid.

There are several functions in MATLAB that can be used to generate numerical grids. For numerical grids in one dimension, it is sufficient to use the colon operator or the `linspace` function, which you have already used for creating regularly spaced arrays.

In MATLAB, a function $f(x)$ can be represented over an interval by computing its value on a grid. Although the function itself may be continuous, this **discrete** or **discretized** representation is useful for numerical calculations and corresponds to data sets that may be acquired in engineering practice.

An Introduction to MATLAB® Programming and Numerical Methods. http://dx.doi.org/10.1016/B978-0-12-420228-3.00017-8

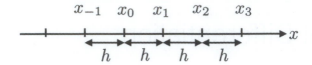

FIGURE 17.1

Numerical grid used to approximate functions.

Specifically, the function value may only be known at discrete points. For example, a temperature sensor may deliver temperature versus time pairs at regular time intervals. Although temperature is a smooth and continuous function of time, the sensor only provides values at discrete time intervals, and in this particular case, the underlying function would not even be known.

Whether f is an analytic function or a discrete representation of one, we would like to derive methods of approximating the derivative of f over a numerical grid and determine their accuracy.

17.2 Approximating Derivatives with Taylor Series

To derive an approximation for the derivative of f, we return to Taylor series. For an arbitrary function $f(x)$ the Taylor series of f around $a = x_j$ is

$$f(x) = \frac{f(x_j)(x - x_j)^0}{0!} + \frac{f'(x_j)(x - x_j)^1}{1!} + \frac{f''(x_j)(x - x_j)^2}{2!} + \frac{f'''(x_j)(x - x_j)^3}{3!} + \cdots.$$

If x is on a grid of points with spacing h, we can compute the Taylor series at $x = x_{j+1}$ to get

$$f(x_{j+1}) = \frac{f(x_j)(x_{j+1} - x_j)^0}{0!} + \frac{f'(x_j)(x_{j+1} - x_j)^1}{1!} + \frac{f''(x_j)(x_{j+1} - x_j)^2}{2!} + \frac{f'''(x_j)(x_{j+1} - x_j)^3}{3!} + \cdots.$$

Substituting $h = x_{j+1} - x_j$ and solving for $f'(x_j)$ gives the equation

$$f'(x_j) = \frac{f(x_{j+1}) - f(x_j)}{h} + \left(-\frac{f''(x_j)h}{2!} - \frac{f'''(x_j)h^2}{3!} - \cdots \right).$$

The terms that are in parentheses, $-\frac{f''(x_j)h}{2!} - \frac{f'''(x_j)h^2}{3!} - \cdots$, are called **higher order terms** of h. The higher order terms can be rewritten as

$$-\frac{f''(x_j)h}{2!} - \frac{f'''(x_j)h^2}{3!} - \cdots = h(\alpha + \epsilon(h)),$$

where α is some constant, and $\epsilon(h)$ is a function of h that goes to zero as h goes to 0. You can verify with some algebra that this is true. We use the abbreviation "$O(h)$" for $h(\alpha + \epsilon(h))$, and in general, we use the abbreviation "$O(h^p)$" to denote $h^p(\alpha + \epsilon(h))$.

Substituting $O(h)$ into the previous equations gives

$$f'(x_j) = \frac{f(x_{j+1}) - f(x_j)}{h} + O(h).$$

This gives the **forward difference** formula for approximating derivatives as

$$f'(x_j) \approx \frac{f(x_{j+1}) - f(x_j)}{h},$$

and we say this formula is $O(h)$.

Here, $O(h)$ describes the **accuracy** of the forward difference formula for approximating derivatives. For an approximation that is $O(h^p)$, we say that p is the **order** of the accuracy of the approximation. With few exceptions, higher order accuracy is better than lower order. To illustrate this point, assume $q < p$. Then as the spacing, $h > 0$, goes to 0, h^p goes to 0 faster than h^q. Therefore as h goes to 0, an approximation of a value that is $O(h^p)$ gets closer to the true value faster than one that is $O(h^q)$.

By computing the Taylor series around $a = x_j$ at $x = x_{j-1}$ and again solving for $f'(x_j)$, we get the **backward difference** formula

$$f'(x_j) \approx \frac{f(x_j) - f(x_{j-1})}{h},$$

which is also $O(h)$. You should try to verify this result on your own.

Intuitively, the forward and backward difference formulas for the derivative at x_j are just the slopes between the point at x_j and the points x_{j+1} and x_{j-1}, respectively.

We can construct an improved approximation of the derivative by clever manipulation of Taylor series terms taken at different points. To illustrate, we can compute the Taylor series around $a = x_j$ at both x_{j+1} and x_{j-1}. Written out, these equations are

$$f(x_{j+1}) = f(x_j) + f'(x_j)h + \frac{1}{2}f''(x_j)h^2 + \frac{1}{6}f'''(x_j)h^3 + \cdots$$

and

$$f(x_{j-1}) = f(x_j) - f'(x_j)h + \frac{1}{2}f''(x_j)h^2 - \frac{1}{6}f'''(x_j)h^3 + \cdots .$$

Subtracting the formulas above gives

$$f(x_{j+1}) - f(x_{j-1}) = 2f'(x_j) + \frac{2}{3}f'''(x_j)h^3 + \cdots ,$$

which when solved for $f'(x_j)$ gives the **central difference** formula

$$f'(x_j) \approx \frac{f(x_{j+1}) - f(x_{j-1})}{2h}.$$

Because of how we subtracted the two equations, the h terms canceled out; therefore, the central difference formula is $O(h^2)$, even though it requires the same amount of computational effort as the forward and backward difference formulas! Thus the central difference formula gets an extra order

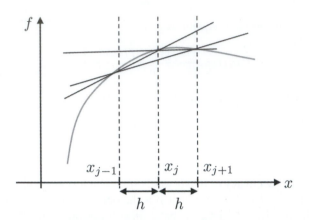

FIGURE 17.2

Illustration of the forward difference (line joining (x_j, f_j) and (x_{j+1}, f_{j+1})), the backward difference (line joining (x_j, f_j), (x_{j-1}, f_{j-1})), and the central difference (line joining (x_{j-1}, f_{j-1}) and (x_{j+1}, f_{j+1})). Note the difference in slopes depending on the method used.

of accuracy for free. In general, formulas that utilize symmetric points around x_j, for example x_{j-1} and x_{j+1}, have better accuracy than asymmetric ones, such as the forward and background difference formulas.

Figure 17.2 shows the forward, backward, and central difference approximation of the derivative of a function f. As can be seen, the difference in the value of the slope can be significantly different based on the size of the step h and the nature of the function.

TRY IT! Take the Taylor series of f around $a = x_j$ and compute the series at $x = x_{j-2}, x_{j-1}, x_{j+1}, x_{j+2}$. Show that the resulting equations can be combined to form an approximation for $f'(x_j)$ that is $O(h^4)$.

First, compute the Taylor series at the specified points.

$$f(x_{j-2}) = f(x_j) - 2hf'(x_j) + \frac{4h^2 f''(x_j)}{2} - \frac{8h^3 f'''(x_j)}{6} + \frac{16h^4 f''''(x_j)}{24} - \frac{32h^5 f'''''(x_j)}{120} + \cdots$$

$$f(x_{j-1}) = f(x_j) - hf'(x_j) + \frac{h^2 f''(x_j)}{2} - \frac{h^3 f'''(x_j)}{6} + \frac{h^4 f''''(x_j)}{24} - \frac{h^5 f'''''(x_j)}{120} + \cdots$$

$$f(x_{j+1}) = f(x_j) + hf'(x_j) + \frac{h^2 f''(x_j)}{2} + \frac{h^3 f'''(x_j)}{6} + \frac{h^4 f''''(x_j)}{24} + \frac{h^5 f'''''(x_j)}{120} + \cdots$$

$$f(x_{j+2}) = f(x_j) + 2hf'(x_j) + \frac{4h^2 f''(x_j)}{2} + \frac{8h^3 f'''(x_j)}{6} + \frac{16h^4 f''''(x_j)}{24} + \frac{32h^5 f'''''(x_j)}{120} + \cdots$$

To get the h^2, h^3, and h^4 terms to cancel out, we can compute

$$f(x_{j-2}) - 8f(x_{j-1}) + 8f(x_{j-1}) - f(x_{j+2}) = 12hf'(x_j) - \frac{48h^5 f'''''(x_j)}{120},$$

which can be rearranged to

$$f'(x_j) = \frac{f(x_{j-2}) - 8f(x_{j-1}) + 8f(x_{j-1}) - f(x_{j+2})}{12h} + O(h^4).$$

This formula is a better approximation for the derivative at x_j than the central difference formula, but requires twice as many calculations.

TIP! MATLAB has a command that can be used to compute finite differences directly: for a vector f, the command d=diff(f) produces an array d in which the entries are the differences of the adjacent elements in the initial array f. In other words d(i) = f(i+1) - f(i).

WARNING! When using the command diff, the size of the output is one less than the size of the input since it needs two arguments to produce a difference.

EXAMPLE: Consider the function $f(x) = \cos(x)$. We know the derivative of $\cos(x)$ is $-\sin(x)$. Although in practice we may not know the underlying function we are finding the derivative for, we use the simple example to illustrate the aforementioned numerical differentiation methods and their accuracy. The following code computes the derivatives numerically.

```
clc; clear all; close all; % start clean

h=0.1; % step size
x = 0:h:2*pi; % define grid
y = cos(x); % compute function

forward_diff    = diff(y)/h; % compute vector of forward differences
x_diff          = x(1:length(x)−1); % compute corresponding grid
exact_solution  = −sin(x_diff); % compute exact solution

% Plot solution
plot(x_diff,forward_diff,'LineStyle','−'); hold on
plot(x_diff,exact_solution)

% Compute max error between numerical derivative and exact solution
max_error       = max(abs(exact_solution−forward_diff))
```

This code produces the following output.

```
max_error =
   0.049984407218554
```

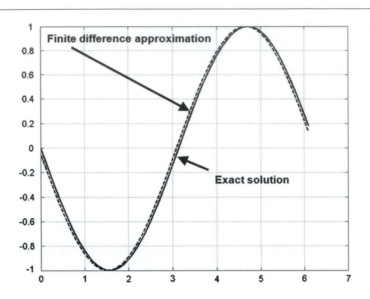

FIGURE 17.3

Comparison of the numerical evaluation of the explicit formula for the derivative of `cos` and of the derivative of `cos` obtained by the forward difference formula.

As Figure 17.3 shows, there is a small offset between the two curves, which results from the numerical error in the evaluation of the numerical derivatives. The maximal error between the two numerical results is of the order 0.05 and expected to decrease with the size of the step.

As illustrated in the previous example, the finite difference scheme contains a numerical error due to the approximation of the derivative. This difference decreases with the size of the discretization step, which is illustrated in the following example.

EXAMPLE: The following code computes the numerical derivative of $f(x) = \cos(x)$ using the forward difference formula for decreasing step sizes, h. It then plots the maximum error between the approximated derivative and the true derivative versus h (Figure 17.4).

```
clc; clear all; close all % start clean

h = 1; % define step size
iterations = 20; % define number of iterations to perform

for i=1:iterations

        h = h/2; % halve the step size
        step_size(i) = h; % store this step size
```

```
        x = 0:h:2*pi; % compute new grid
        y = cos(x); % compute function value at grid

    forward_diff      = diff(y)/h; % compute vector of forward differences
    x_diff            = x(1:length(x)-1); % compute corresponding grid
    exact_solution    = -sin(x_diff); % compute exact solution

    % Compute max error between numerical derivative and exact solution
        max_error(i)     = max(abs(exact_solution-forward_diff));
end

% produce log-log plot of max error versus step size
loglog(step_size,max_error,'LineWidth',2,'LineStyle','v'); grid on
```

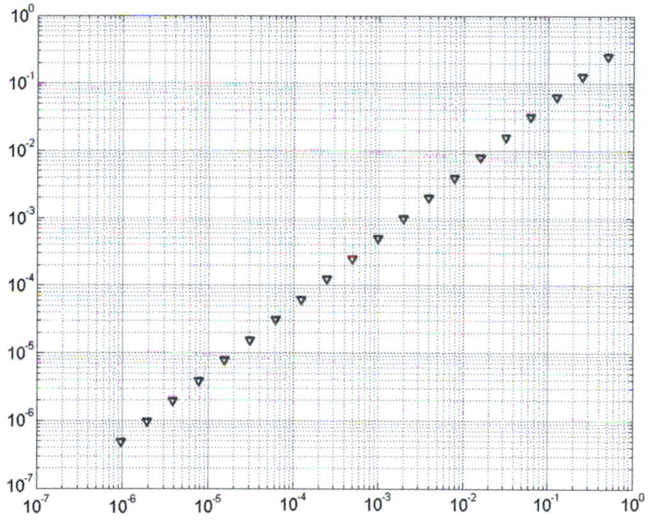

FIGURE 17.4

Maximum error between the numerical evaluation of the explicit formula for the derivative of cos and the derivative of cos obtained by forward finite differencing.

The slope of the line in log-log space is 1; therefore, the error is proportional to h^1, which means that, as expected, the forward difference formula is $O(h)$.

17.3 Approximations of Higher Order Derivatives

It also possible to use Taylor series to approximate higher order derivatives (e.g., $f''(x_j)$, $f'''(x_j)$, etc.). For example, taking the Taylor series around $a = x_j$ and then computing it at $x = x_{j-1}$ and x_{j+1} gives

$$f(x_{j-1}) = f(x_j) - hf'(x_j) + \frac{h^2 f''(x_j)}{2} - \frac{h^3 f'''(x_j)}{6} + \cdots$$

and

$$f(x_{j+1}) = f(x_j) + hf'(x_j) + \frac{h^2 f''(x_j)}{2} + \frac{h^3 f'''(x_j)}{6} + \cdots.$$

If we add these two equations together, we get

$$f(x_{j-1}) + f(x_{j+1}) = 2f(x_j) + h^2 f''(x_j) + \frac{h^4 f''''(x_j)}{24} + \cdots,$$

and with some rearrangement gives the approximation

$$f''(x_j) \approx \frac{f(x_{j+1}) - 2f(x_j) + f(x_{j-1})}{h^2},$$

and is $O(h^2)$.

17.4 Numerical Differentiation with Noise

As stated earlier, sometimes f is given as a vector where f is the corresponding function value for independent data values in another vector x, which is gridded. Sometimes data can be contaminated with **noise**, meaning its value is off by a small amount from what it would be if it were computed from a pure mathematical function. This can often occur in engineering due to inaccuracies in measurement devices or the data itself can be slightly modified by perturbations outside the system of interest. For example, you may be trying to listen to your friend talk in a crowded room. The signal f might be the intensity and tonal values in your friend's speech. However, because the room is crowded, noise from other conversations are heard along with your friend's speech, and he becomes difficult to understand.

To illustrate this point, we numerically compute the derivative of a simple cosine wave corrupted by a small sin wave. Consider the following two functions:

$$f(x) = \cos(x)$$

and

$$f_{\epsilon,\omega}(x) = \cos(x) + \epsilon \sin(\omega x)$$

where $0 < \epsilon \ll 1$ is a very small number and ω is a large number. When ϵ is small, it is clear that $f \simeq f_{\epsilon,\omega}$. To illustrate this point, we plot $f_{\epsilon,\omega}(x)$ for $\epsilon = 0.01$ and $\omega = 100$, and we can see it is very close to $f(x)$, as shown in Figure 17.5.

The derivatives of our two test functions are

$$f'(x) = -\sin(x)$$

and

$$f'_{\epsilon,\omega}(x) = -\sin(x) + \epsilon\omega \cos(\omega x).$$

Since $\epsilon\omega$ may not be small when ω is large, the contribution of the noise to the derivative may not be small. As a result, the derivative (analytic and numerical) may not be usable. For instance, Figure 17.6 shows $f'(x)$ and $f'_{\epsilon,\omega}(x)$ for $\epsilon = 0.01$ and $\omega = 100$.

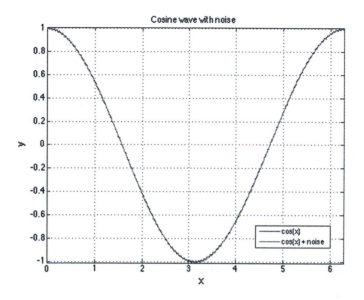

FIGURE 17.5

Cosine wave contaminated with a small amount of noise. The noise is hardly visible, but it will be shown that it has drastic consequences for the derivative.

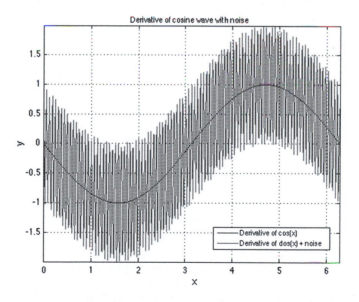

FIGURE 17.6

Although the noise is hardly visible on the function itself, the noise makes the derivative completely unusable.

Summary

1. Because explicit derivation of functions is sometimes cumbersome for engineering applications, numerical approaches can be preferable.
2. Numerical approximation of derivatives can be done using a grid on which the derivative is approximated by finite differences.
3. Finite differences approximate the derivative by ratios of differences in the function value over small intervals.
4. Finite difference schemes have different approximation orders depending on the method used.
5. There are issues with finite differences for approximation of derivatives when the data is noisy.

Vocabulary

accuracy	discretize	numerical grid
backward difference	forward difference	order
central difference	higher order terms	spacing
discrete	noise	step size

Functions and Operators

```
diff
```

Problems

⒨ 1. Write a function with header [df, X] = myDerCalc(f, a, b, N, option), where f is a function handle, a and b are scalars such that a < b, N is an integer bigger than 10, and option is the string 'forward', 'backward', or 'central'. Let x be an array starting at a, ending at b, containing N evenly spaced elements, and let y be the array f(x). The output argument, df, should be the numerical derivatives computed for x and y according to the method defined by the input argument, option. The output argument X should be an array the same size as df containing the points in x for which df is valid. Specifically, the forward difference method "loses" the last point, the backward difference method loses the first point, and the central difference method loses the first and last points.

⒨ 2. Write a function with header [dy, X] = myNumDiff(f, a, b, n, option), where f is a handle to a function. The function myNumDiff should compute the derivative of f numerical for n evenly spaced points starting at a and ending at b according to the method defined by option. The input argument option is one of the following strings: 'forward', 'backward', 'central'. Note that for the forward and backward method, the output argument, dy, should be $1 \times (n-1)$, and for the central difference method dy should be $1 \times (n-2)$.

The function should also output a row vector X that is the same size as dy and denotes the x-values for which dy is valid.

Test Cases:

```
>> x = linspace(0, 2*pi,100);
>> f = @sin;
>> [dyf, Xf] = myNumDiff(f, 0, 2*pi, 10, 'forward');
>> [dyb, Xb] = myNumDiff(f, 0, 2*pi, 10, 'backward');
>> [dyc, Xc] = myNumDiff(f, 0, 2*pi, 10, 'central');
>> plot(x, cos(x), Xf, dyf, Xb, dyb, Xc, dyc)
>> title('Analytic and Numerical Derivatives of Sine')
>> xlabel('x')
>> ylabel('y')
>> grid on
>> legend('analytic', 'forward', 'backward', 'central')
```

```
>> x = linspace(0, pi, 1000);
>> f = @(x) sin(exp(x));
>> [dy10, X10] = myNumDiff(f, 0, pi, 10, 'central');
>> [dy20, X20] = myNumDiff(f, 0, pi, 20, 'central');
>> [dy100, X100] = myNumDiff(f, 0, pi, 100, 'central');
>> plot(x, cos(exp(x)).*exp(x), X10, dy10, X20, dy20, X100, dy100)
>> xlabel('x')
>> ylabel('y')
>> title('Numerical Derivative of f(x) = sin(e^{x})')
>> legend('Analytic', '10 points', '20 points', '100 points')
```

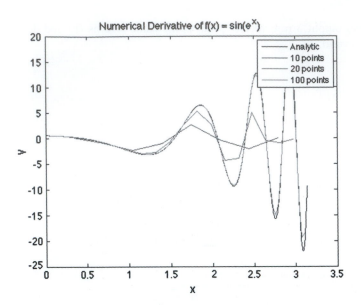

⌊m⌋ 3. Write a function with header [dy, X] = myNumDiffwSmoothing(x, y, n), where x and y are row vectors of the same length, and n is a strictly positive scalar. The function should first create a vector of "smoothed" y data points where ySmooth(i) = mean(y(i-n:i+n)). The function should then compute dy, the derivative of the smoothed y-vector using the central difference method. The function should also output a row vector X that is the same size as dy and denotes the x-values for which dy is valid.

Assume that the data contained in x is in ascending order with no duplicate entries. However, it is possible that the elements of x will not be evenly spaced. Note that the output dy will have $2n + 2$ fewer points than y. Assume that the length of y is much bigger than $2n + 2$.

Test Cases:

```
>> x = linspace(0,2*pi,100);
>> y = sin(x) + randn(size(x))/100;
>> [dy,X] = myNumDiffwSmoothing(x, y, 4);
>> subplot(2,1,1)
>> plot(x, y);
>> title('Noisy Sine function')
>> xlabel('x')
>> ylabel('y')
>> grid on
>> subplot(2,1,2)
>> plot(x,cos(x),'b', x(1:end-1), (y(2:end) - y(1:end-1))/(x(2)-x(1)), 'r', X, dy, 'g')
>> title('Analytic Derivative and Smoothed Derivative')
>> xlabel('x')
>> ylabel('y')
>> grid on
>> legend('cosine', 'unsmoothed forward diff', 'smoothed')
```

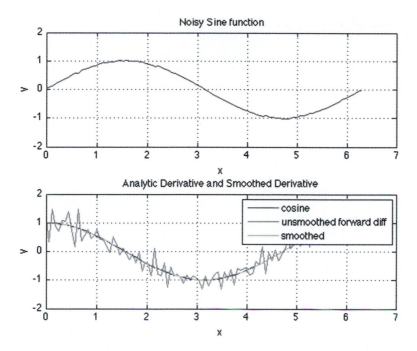

Noisy Sine function

Analytic Derivative and Smoothed Derivative

4. Use Taylor series to show the following approximations and their accuracy.

$$f''(x_j) = \frac{-f(x_{j+3}) + 4f(x_{j+2}) - 5f(x_{j+1}) + 2f(x_j)}{h^2} + O(h^2),$$

$$f'''(x_j) = \frac{f(x_{j+3}) - 3f(x_{j+2}) + 3f(x_{j+1}) - f(x_j)}{h^3} + O(h).$$

Numerical Integration

CHAPTER OUTLINE

Motivation

The integral of a function is normally described as the "area under the curve." In engineering, the integral has many applications for modeling, predicting, and understanding physical systems. However in practice, finding an exact solution for the integral of a function is difficult or impossible.

This chapter describes several methods of numerically integrating functions. By the end of this chapter, you should understand these methods, how they are derived, their geometric interpretation, and their accuracy.

18.1 Numerical Integration Problem Statement

Given a function $f(x)$, we want to approximate the integral of $f(x)$ over the total **interval**, $[a, b]$. Figure 18.1 illustrates this area. To accomplish this goal, we assume that the interval has been discretized into a numeral grid, x, consisting of $n + 1$ points with spacing, $h = \frac{b-a}{n}$. Here, we denote each point in x by x_i, where $x_0 = a$ and $x_n = b$. Note: There are $n + 1$ grid points because the count starts at x_0. We also assume we have a function, $f(x)$, that can be computed for any of the grid points, or that we have been given the function implicitly as $f(x_i)$. The interval $[x_i, x_{i+1}]$ is referred to as a **subinterval**.

The following sections give some of the most common methods of approximating $\int_a^b f(x)dx$. Each method approximates the area under $f(x)$ for each subinterval by a shape for which it is easy to compute the exact area, and then sums the area contributions of every subinterval.

An Introduction to MATLAB® Programming and Numerical Methods. http://dx.doi.org/10.1016/B978-0-12-420228-3.00018-X

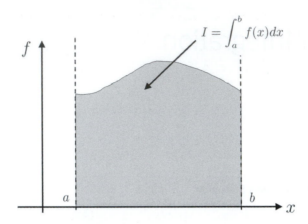

FIGURE 18.1

Illustration of the integral. The integral from *a* to *b* of the function *f* is the area below the curve (shaded in grey).

18.2 Riemann's Integral

The simplest method for approximating integrals is by summing the area of rectangles that are defined for each subinterval. The width of the rectangle is $x_{i+1} - x_i = h$, and the height is defined by a function value $f(x)$ for some x in the subinterval. An obvious choice for the height is the function value at the left endpoint, x_i, or the right endpoint, x_{i+1}, because these values can be used even if the function itself is not known. This method gives the **Riemann Integral** approximation, which is

$$\int_a^b f(x)dx \approx \sum_{i=0}^{n-1} hf(x_i),$$

or

$$\int_a^b f(x)dx \approx \sum_{i=1}^{n} hf(x_i),$$

depending on whether the left or right endpoint is chosen.

As with numerical differentiation, we want to characterize how the accuracy improves as h gets small. To determine this characterizing, we first rewrite the integral of $f(x)$ over an arbitrary subinterval in terms of the Taylor series. The Taylor series of $f(x)$ around $a = x_i$ is

$$f(x) = f(x_i) + f'(x_i)(x - x_i) + \cdots$$

Thus

$$\int_{x_i}^{x_{i+1}} f(x)dx = \int_{x_i}^{x_{i+1}} \left(f(x_i) + f'(x_i)(x - x_i) + \cdots \right) dx$$

by substitution of the Taylor series for the function. Since the integral distributes, we can rearrange the right side into the following form:

$$\int_{x_i}^{x_{i+1}} f(x_i)dx + \int_{x_i}^{x_{i+1}} f'(x_i)(x - x_i)dx + \cdots .$$

Solving each integral separately results in the approximation

$$\int_{x_i}^{x_{i+1}} f(x)dx = hf(x_i) + \frac{h^2}{2}f'(x_i) + O(h^3),$$

which is just

$$\int_{x_i}^{x_{i+1}} f(x)dx = hf(x_i) + O(h^2).$$

Since the $hf(x_i)$ term is our Riemann integral approximation for a single subinterval, the Riemann integral approximation over a single interval is $O(h^2)$.

If we sum the $O(h^2)$ error over the entire Riemann sum, we get $nO(h^2)$. The relationship between n and h is

$$h = \frac{b-a}{n},$$

and so our total error becomes $\frac{b-a}{h}O(h^2) = O(h)$ over the whole interval. Thus the overall accuracy is $O(h)$.

The **Midpoint Rule** takes the rectangle height of the rectangle at each subinterval to be the function value at the midpoint between x_i and x_{i+1}, which for compactness we denote by $y_i = \frac{x_{i+1}+x_i}{2}$. The Midpoint Rule says

$$\int_a^b f(x)dx \approx \sum_{i=0}^{n-1} hf(y_i).$$

Similarly to the Riemann integral, we take the Taylor series of $f(x)$ around y_i, which is

$$f(x) = f(y_i) + f'(y_i)(x - y_i) + \frac{f''(y_i)(x - y_i)^2}{2!} + \cdots$$

Then the integral over a subinterval is

$$\int_{x_i}^{x_{i+1}} f(x)dx = \int_{x_i}^{x_{i+1}} \left(f(y_i) + f'(y_i)(x - y_i) + \frac{f''(y_i)(x - y_i)^2}{2!} + \cdots \right) dx,$$

which distributes to

$$\int_{x_i}^{x_{i+1}} f(x)dx = \int_{x_i}^{x_{i+1}} f(y_i)dx + \int_{x_i}^{x_{i+1}} f'(y_i)(x - y_i)dx + \int_{x_i}^{x_{i+1}} \frac{f''(y_i)(x - y_i)^2}{2!}dx + \cdots .$$

Recognizing that since x_i and x_{i+1} are symmetric around y_i, then $\int_{x_i}^{x_{i+1}} f'(y_i)(x - y_i)dx = 0$. This is true for the integral of $(x - y_i)^p$ for any odd p. For the integral of $(x - y_i)^p$ and with p even, it suffices to say that $\int_{x_i}^{x_{i+1}} (x - y_i)^p dx = \int_{-\frac{h}{2}}^{\frac{h}{2}} x^p dx$, which will result in some multiple of h^{p+1} with no lower order powers of h.

Utilizing these facts reduces the expression for the integral of $f(x)$ to

$$\int_{x_i}^{x_{i+1}} f(x)dx = hf(y_i) + O(h^3).$$

Since $hf(y_i)$ is the approximation of the integral over the subinterval, the Midpoint Rule is $O(h^3)$ for one subinterval, and using similar arguments as for the Riemann Integral, is $O(h^2)$ over the whole interval. Since the Midpoint Rule requires the same number of calculations as the Riemann Integral, we essentially get an extra order of accuracy for free! However, if $f(x_i)$ is given in the form of data points, then we will not be able to compute $f(y_i)$ for this integration scheme.

TRY IT! Use the left Riemann Integral, right Riemann Integral, and Midpoint Rule to approximate $\int_0^\pi \sin(x)dx$ with 11 evenly spaced grid points over the whole interval. Compare this value to the exact value of 2.

```
a = 0;
b = pi;
n = 11;
h = (b−a)/(n−1);
x = linspace(a,b,n);
f = sin(x);

I_riemannL = h*sum(f(1:n−1))
err_riemannL = 2 − I_riemannL

I_riemannR = h*sum(f(2:n))
err_riemannR = 2 − I_riemannR

I_mid = h*sum(sin((x(1:n−1) + x(2:n))/2))
err_mid = 2 − I_mid
```

The previous code produces the following output.

```
I_riemannL =
    1.983523537509455
err_riemannL =
    0.016476462490545
```

```
I_riemannR =
    1.983523537509455
err_riemannR =
    0.016476462490545

I_mid =
    2.008248407907975
err_mid =
   -0.008248407907975
```

18.3 Trapezoid Rule

The **Trapezoid Rule** fits a trapezoid into each subinterval and sums the areas of the trapezoid to approximate the total integral. This approximation for the integral to an arbitrary function is shown in Figure 18.2. For each subinterval, the Trapezoid Rule computes the area of a trapezoid with corners at $(x_i, 0)$, $(x_{i+1}, 0)$, $(x_i, f(x_i))$, and $(x_{i+1}, f(x_{i+1}))$, which is $h\frac{f(x_i)+f(x_{i+1})}{2}$. Thus, the Trapezoid Rule approximates integrals according to the expression

$$\int_a^b f(x)dx \approx \sum_{i=0}^{n-1} h\frac{f(x_i) + f(x_{i+1})}{2}.$$

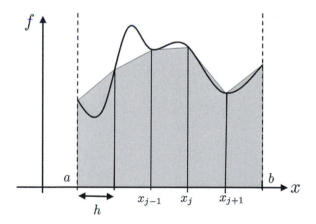

FIGURE 18.2

Illustration of the trapezoid integral procedure. The area below the curve is approximated by a summation of trapezoids that approximate the function.

TIP! You may notice that the Trapezoid Rule "double-counts" most of the terms in the series. To illustrate this fact, consider the expansion of the Trapezoid Rule:

$$\sum_{i=0}^{n-1} h \frac{f(x_i) + f(x_{i+1})}{2} = \frac{h}{2}\left[(f(x_0) + f(x_1)) + (f(x_1) + f(x_2)) + (f(x_2)\right.$$

$$\left. + f(x_3)) + \cdots + (f(x_{n-1}) + f(x_n))\right].$$

Computationally, this is many extra additions and calls to $f(x)$ than is really necessary. We can be more computationally efficient using the following expression.

$$\int_a^b f(x)dx \approx \frac{h}{2}\left(f(x_0) + 2\left(\sum_{i=1}^{n-1} f(x_i)\right) + f(x_n)\right).$$

To determine the accuracy of the Trapezoid Rule approximation, we first take Taylor series expansion of $f(x)$ around $y_i = \frac{x_{i+1}+x_i}{2}$, which is the midpoint between x_i and x_{i+1}. This Taylor series expansion is

$$f(x) = f(y_i) + f'(y_i)(x - y_i) + \frac{f''(y_i)(x - y_i)^2}{2!} + \cdots$$

Computing the Taylor series at x_i and x_{i+1} and noting that $x_i - y_i = -\frac{h}{2}$ and $x_{i+1} - y_i = \frac{h}{2}$, results in the following expressions:

$$f(x_i) = f(y_i) - \frac{hf'(y_i)}{2} + \frac{h^2 f''(y_i)}{8} - \cdots$$

and

$$f(x_{i+1}) = f(y_i) + \frac{hf'(y_i)}{2} + \frac{h^2 f''(y_i)}{8} + \cdots .$$

Taking the average of these two expressions results in the new expression,

$$\frac{f(x_{i+1}) + f(x_i)}{2} = f(y_i) + O(h^2).$$

Solving this expression for $f(y_i)$ yields

$$f(y_i) = \frac{f(x_{i+1}) + f(x_i)}{2} + O(h^2).$$

Now returning to the Taylor expansion for $f(x)$, the integral of $f(x)$ over a subinterval is

$$\int_{x_i}^{x_{i+1}} f(x)dx = \int_{x_i}^{x_{i+1}} \left(f(y_i) + f'(y_i)(x - y_i) + \frac{f''(y_i)(x - y_i)^2}{2!} + \cdots\right) dx.$$

Distributing the integral results in the expression

$$\int_{x_i}^{x_{i+1}} f(x)dx = \int_{x_i}^{x_{i+1}} f(y_i)dx + \int_{x_i}^{x_{i+1}} f'(y_i)(x - y_i)dx + \int_{x_i}^{x_{i+1}} \frac{f''(y_i)(x - y_i)^2}{2!}dx + \cdots$$

Now since x_i and x_{i+1} are symmetric around y_i, the integrals of the odd powers of $(x - y_i)^p$ disappear and the even powers resolve to a multiple h^{p+1}.

$$\int_{x_i}^{x_{i+1}} f(x)dx = hf(y_i) + O(h^3).$$

Now if we substitute $f(y_i)$ with the expression derived explicitly in terms of $f(x_i)$ and $f(x_{i+1})$, we get

$$\int_{x_i}^{x_{i+1}} f(x)dx = h\left(\frac{f(x_{i+1}) + f(x_i)}{2} + O(h^2)\right) + O(h^3),$$

which is equivalent to

$$h\left(\frac{f(x_{i+1}) + f(x_i)}{2}\right) + hO(h^2) + O(h^3)$$

and therefore,

$$\int_{x_i}^{x_{i+1}} f(x)dx = h\left(\frac{f(x_{i+1}) + f(x_i)}{2}\right) + O(h^3).$$

Since $\frac{h}{2}(f(x_{i+1}) + f(x_i))$ is the Trapezoid Rule approximation for the integral over the subinterval, it is $O(h^3)$ for a single subinterval and $O(h^2)$ over the whole interval.

TRY IT! Use the Trapezoid Rule to approximate $\int_0^\pi \sin(x)dx$ with 11 evenly spaced grid points over the whole interval. Compare this value to the exact value of 2.

```
format long
a = 0;
b = pi;
n = 11;
h = (b-a)/(n-1);
x = linspace(a,b,n);
f = sin(x);
I_trap = (h/2)*(f(1) + 2*sum(f(2:n-1)) + f(n))
err_trap = 2 - I_trap
```

The previous code produces the following output.

```
I_trap =
   1.983523537509455

err_trap =
   0.016476462490545
```

18.4 Simpson's Rule

Consider *two* consecutive subintervals, $[x_{i-1}, x_i]$ and $[x_i, x_{i+1}]$. **Simpson's Rule** approximates the area under $f(x)$ over these two subintervals by fitting a quadratic polynomial through the points $(x_{i-1}, f(x_{i-1}))$, $(x_i, f(x_i))$, and $(x_{i+1}, f(x_{i+1}))$, which is a unique polynomial, and then integrating the quadratic exactly. Figure 18.3 shows this integral approximation for an arbitrary function.

First, we construct the quadratic polynomial approximation of the function over the two subintervals. The easiest way to do this is to use Lagrange polynomials, which was discussed in Chapter 14. By applying the formula for constructing Lagrange polynomials we get the polynomial

$$P_i(x) = f(x_{i-1}) \frac{(x - x_i)(x - x_{i+1})}{(x_{i-1} - x_i)(x_{i-1} - x_{i+1})} + f(x_i) \frac{(x - x_{i-1})(x - x_{i+1})}{(x_i - x_{i-1})(x_i - x_{i+1})}$$
$$+ f(x_{i+1}) \frac{(x - x_{i-1})(x - x_i)}{(x_{i+1} - x_{i-1})(x_{i+1} - x_i)},$$

and with substitutions for h results in

$$P_i(x) = \frac{f(x_{i-1})}{2h^2}(x - x_i)(x - x_{i+1}) - \frac{f(x_i)}{h^2}(x - x_{i-1})(x - x_{i+1}) + \frac{f(x_{i+1})}{2h^2}(x - x_{i-1})(x - x_i).$$

You can confirm that the polynomial intersects the desired points. With some algebra and manipulation, the integral of $P_i(x)$ over the two subintervals is

$$\int_{x_{i-1}}^{x_{i+1}} P_i(x)dx = \frac{h}{3}(f(x_{i-1}) + 4f(x_i) + f(x_{i+1})).$$

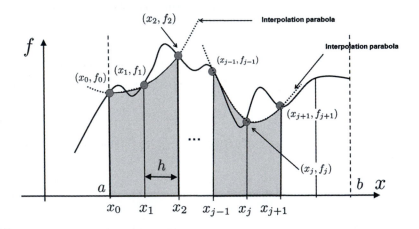

FIGURE 18.3

Illustration of the Simpson integral formula. Discretization points are grouped by three, and a parabola is fit between the three points. This can be done by a typical interpolation polynomial. The area under the curve is approximated by the area under the parabola.

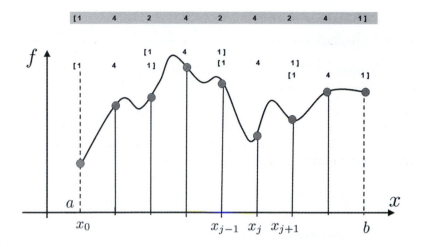

FIGURE 18.4

Illustration of the accounting procedure to approximate the function f by the Simpson rule for the entire interval [a, b].

To approximate the integral over (a, b), we must sum the integrals of $P_i(x)$ over every *two* subintervals since $P_i(x)$ spans two subintervals. Substituting $\frac{h}{3}(f(x_{i-1}) + 4f(x_i) + f(x_{i+1}))$ for the integral of $P_i(x)$ and regrouping the terms for efficiency leads to the formula

$$\int_a^b f(x)dx \approx \frac{h}{3}\left[f(x_0) + 4\left(\sum_{i=1, i \text{ odd}}^{n-1} f(x_i) \right) + 2\left(\sum_{i=2, i \text{ even}}^{n-2} f(x_i) \right) + f(x_n) \right].$$

This regrouping is illustrated in Figure 18.4.

> **WARNING!** Note that to use Simpson's Rule, you *must* have an even number of intervals and, therefore, an odd number of grid points.

To compute the accuracy of the Simpson's Rule, we take the Taylor series approximation of $f(x)$ around x_i, which is

$$f(x) = f(x_i) + f'(x_i)(x - x_i) + \frac{f''(x_i)(x - x_i)^2}{2!} + \frac{f'''(x_i)(x - x_i)^3}{3!} + \frac{f''''(x_i)(x - x_i)^4}{4!} + \cdots$$

Computing the Taylor series at x_{i-1} and x_{i+1} and substituting for h where appropriate gives the expressions

$$f(x_{i-1}) = f(x_i) - hf'(x_i) + \frac{h^2 f''(x_i)}{2!} - \frac{h^3 f'''(x_i)}{3!} + \frac{h^4 f''''(x_i)}{4!} - \cdots$$

and

$$f(x_{i+1}) = f(x_i) + hf'(x_i) + \frac{h^2 f''(x_i)}{2!} + \frac{h^3 f'''(x_i)}{3!} + \frac{h^4 f''''(x_i)}{4!} + \cdots$$

Now consider the expression $\frac{f(x_{i-1})+4f(x_i)+f(x_{i+1})}{6}$. Substituting the Taylor series for the respective numerator values produces the equation

$$\frac{f(x_{i-1}) + 4f(x_i) + f(x_{i+1})}{6} = f(x_i) + \frac{h^2}{6} f''(x_i) + \frac{h^4}{72} f''''(x_i) + \cdots$$

Note that the odd terms cancel out. This implies

$$f(x_i) = \frac{f(x_{i-1}) + 4f(x_i) + f(x_{i+1})}{6} - \frac{h^2}{6} f''(x_i) + O(h^4).$$

By substitution of the Taylor series for $f(x)$, the integral of $f(x)$ over two subintervals is then

$$\int_{x_{i-1}}^{x_{i+1}} f(x)dx = \int_{x_{i-1}}^{x_{i+1}} \left(f(x_i) + f'(x_i)(x - x_i) + \frac{f''(x_i)(x - x_i)^2}{2!} \right.$$
$$\left. + \frac{f'''(x_i)(x - x_i)^3}{3!} + \frac{f''''(x_i)(x - x_i)^4}{4!} + \cdots \right) dx.$$

Again, we distribute the integral and without showing it, we drop the integrals of terms with odd powers because they are zero.

$$\int_{x_{i-1}}^{x_{i+1}} f(x)dx = \int_{x_{i-1}}^{x_{i+1}} f(x_i)dx + \int_{x_{i-1}}^{x_{i+1}} \frac{f''(x_i)(x - x_i)^2}{2!} dx + \int_{x_{i-1}}^{x_{i+1}} \frac{f''''(x_i)(x - x_i)^4}{4!} dx + \cdots.$$

We then carry out the integrations. As will soon be clear, it benefits us to compute the integral of the second term exactly. The resulting equation is

$$\int_{x_{i-1}}^{x_{i+1}} f(x)dx = 2hf(x_i) + \frac{h^3}{3} f''(x_i) + O(h^5).$$

Substituting the expression for $f(x_i)$ derived earlier, the right-hand side becomes

$$2h \left(\frac{f(x_{i-1}) + 4f(x_i) + f(x_{i+1})}{6} - \frac{h^2}{6} f''(x_i) + O(h^4) \right) + \frac{h^3}{3} f''(x_i) + O(h^5),$$

which can be rearranged to

$$\left[\frac{h}{3}(f(x_{i-1}) + 4f(x_i) + f(x_{i+1})) - \frac{h^3}{3} f''(x_i) + O(h^5) \right] + \frac{h^3}{3} f''(x_i) + O(h^5).$$

Canceling and combining the appropriate terms results in the integral expression

$$\int_{x_{i-1}}^{x_{i+1}} f(x)dx = \frac{h}{3}(f(x_{i-1}) + 4f(x_i) + f(x_{i+1})) + O(h^5).$$

Recognizing that $\frac{h}{3}(f(x_{i-1}) + 4f(x_i) + f(x_{i+1}))$ is exactly the Simpson's Rule approximation for the integral over this subinterval, this equation implies that Simpson's Rule is $O(h^5)$ over a subinterval and $O(h^4)$ over the whole interval. Because the h^3 terms cancel out exactly, Simpson's Rule gains another *two* orders of accuracy!

TRY IT! Use Simpson's Rule to approximate $\int_0^\pi \sin(x)dx$ with 11 evenly spaced grid points over the whole interval. Compare this value to the exact value of 2.

```
format long
a = 0;
b = pi;
n = 11;
h = (b-a)/(n-1);
x = linspace(a,b,n);
f = sin(x);

I_simp = (h/3)*(f(1) + 2*sum(f(1:2:n-2)) + 4*sum(f(2:2:n-1)) + f(n))
err_simp = 2 - I_simp
```

Note the change in indices from the formula given in the text. This change was made to accommodate the fact that x_0 in the formula is x(1) in the code. The previous code produces the following output.

```
I_simp =
   2.000109517315004

err_simp =
   -1.095173150043038e-04
```

18.5 Computing Integrals in MATLAB®

MATLAB has several built-in functions for computing integrals. The trapz takes as input arguments a numerical grid x and an array of function values, f, computed on x.

TRY IT! Use the `trapz` function to approximate $\int_0^\pi \sin(x)dx$ for 11 equally spaced points over the whole interval. Compare this value to the one computed in the early example using the Trapezoid Rule.

```
format long
a = 0;
b = pi;
n = 11;
h = (b−a)/(n−1);
x = linspace(a,b,n);
f = sin(x);

I_trapz = trapz(x,f)
I_trap = (h/2)*(f(1) + 2*sum(f(2:n−1)) + f(n))
```

The previous code produces the following output.

```
I_trapz =
    1.983523537509455

I_trap =
    1.983523537509455
```

Sometimes we want to know the approximated cumulative integral. That is, we want to know $F(X) = \int_{x_0}^{X} f(x)dx$. For this purpose, it is useful to use the `cumtrapz` function `cumsum`, which takes the same input arguments as `trapz`.

TRY IT! Use the `cumtrapz` function to approximate the cumulative integral of $f(x) = \sin(x)$ from 0 to π with a discretization step of 0.01. The exact solution of this integral is $F(x) = sin(x)$. Plot the results to compare (Figure 18.5).

```
clc; clear all; close all;

x = 0:0.01:pi;
F_exact = −cos(x);
F_approx = cumtrapz(x,sin(x));

plot(x,F_exact, x, F_approx)
grid on
axis tight
title('F(x) = \int_0^{x} sin(y) dy')
xlabel('x')
ylabel('f(x)')
legend('Exact with Offset', 'Approx')
```

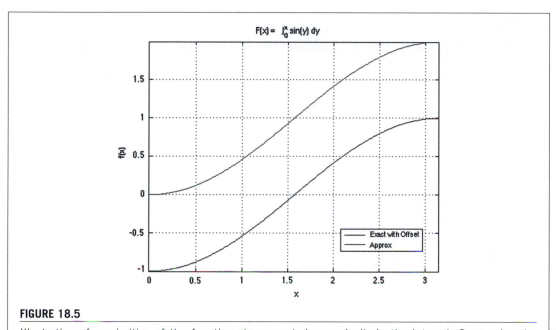

FIGURE 18.5

Illustration of a primitive of the function sin computed numerically in the interval $[0, \pi]$ using the `cumtrapz` function. The result is clearly $-\cos$ with an offset.

The `quad(f,a,b)` function uses a different numerical differentiation scheme to approximate integrals. `quad` integrates the function defined by the function handle, f, from a to b.

TRY IT! Use the `quad` function to compute $\int_0^\pi \sin(x)dx$. Compare your answer with the correct answer of 2.

```
format long
I_quad = quad(@sin, 0, pi)
I_quad =
   1.999999996398431

err_quad = 2 - I_quad
err_quad =
     3.601569042999131e-09
```

Summary

1. Explicit integration of functions is often impossible or inconvenient, and numerical approaches must be used instead.

2. The Riemann Integral, Trapezoid Rule, and Simpson's Rule are common methods of approximating integrals.

3. Each method has an order of accuracy that depends on the approximation of the area below the function.

Vocabulary

Midpoint Rule   Simpson's Rule
Riemann Integral   Trapezoidal Rule

Functions and Operators

cumsum   quad   quadv
cumtrapz   quad2d   sum
dblquad   quadgk   trapz
polyint   quadl   triplequad

Problems

1. Write a function with header `[I] = myIntCalc(f, f0, a, b, N, option)`, where `f` is a function handle, `a` and `b` are scalars such that `a < b`, `N` is a positive integer, and `option` is the string `'rect'`, `'trap'`, or `'simp'`. Let `x` be an array starting at `a`, ending at `b`, and containing `N` evenly spaced elements, and let `y` be the array $f(x)$. The output argument, `I`, should be an approximation to the integral of $f(x)$, with initial condition `f0`, computed according to the input argument, `option`.

2. Write a function with header `[I] = myPolyInt(x, y)`, where `x` and `y` are one-dimensional arrays of the same size, and the elements of `x` are unique and in ascending order. The function `myPolyInt` should (1) compute the Lagrange polynomial going through all the points defined by `x` and `y` and (2) return an approximation to the area under the curve defined by `x` and `y`, `I`, defined as the analytic integral of the Lagrange interpolating polynomial.

3. When will `myPolyInt` work *worse* than the trapezoid method?

4. Write a function with header `[I] = myNumInt(f, a, b, n, option)`, where `I` is the numerical integral of `f`, a function handle, computed on a grid of *n* evenly spaced points starting at `a` and ending at `b`. The integration method used should be one of the following strings defined by option: `'rect'`, `'trap'`, `'simp'`. For the rectangle method, the function value should be taken from the right endpoint of the interval. You may assume that *n* is odd and that `f` is vectorized.

Warning: In the reader, the x subscripts start at x_0 not x_1; take note of this when programming your loops. The odd-even indices will be reversed. Also the n term given in Simpsons Rule denotes the number of subintervals, not the number of points as specified by the input argument, n.

Test Cases:

```
>> format long
>> f = @(x) x.^2;
>> I = myNumInt(f, 0, 1, 3 , 'rect')
I =
   0.625000000000000
>> I = myNumInt(f, 0, 1, 3 , 'trap')
I =
   0.375000000000000
>> I = myNumInt(f, 0, 1, 3 , 'simp')
I =
   0.333333333333333

>> format long
>> f = @(x) exp(x.^2);
>> I = myNumInt(f, -1, 1, 101, 'simp')
I =
   2.925303588392652
>> I = myNumInt(f, -1, 1, 10001, 'simp')
I =
   2.925303491814042
>> I = myNumInt(f, -1, 1, 1000001, 'simp')
I =
   2.925303491736135
```

m 5. A previous chapter demonstrated that some functions can be expressed as an infinite sum of polynomials (i.e. Taylor serie). Other functions, particularly periodic functions, can be written as an infinite sum of sine and cosine waves. For these functions,

$$f(x) = \frac{A_0}{2} + \sum_{n=1}^{\infty} A_n \cos(nx) + B_n \sin(nx)$$

It can be shown that the values of A_n and B_n can be computed using the following formulas:

$$A_n = \frac{1}{\pi} \int_{-\pi}^{\pi} f(x) \cos(nx) \, dx$$

$$B_n = \frac{1}{\pi} \int_{-\pi}^{\pi} f(x) \sin(nx) \, dx$$

Just like Taylor series, functions can be approximated by truncating the Fourier series at some $n = N$. Fourier series can be used to approximate some particularly nasty functions such as the step function, and they form the basis of many engineering applications such as signal processing.

Write a function with header `[An, Bn] = myFourierCoeff(f, n)`, where `f` is a handle to a vectorized function that is 2π-periodic. The function `myFourierCoeff` should compute the n-th Fourier coefficients, A_n and B_n, in the Fourier series for `f` defined by the two formulas given earlier. You should use the `quad` function to perform the integration.

Test Cases: Run this script for the following functions, `f`, and orders, `N` (i.e., you should replace the lines "$f = \cdots$" with the given function and "$N = \cdots$" with order specified in the title of the plots).

```
f = ...
N = ...
x = linspace(-pi, pi, 10000);

[A0, B0] = myFourierCoeff(f, 0);
y = A0*ones(size(x))/2;
for n = 1:N
    [An, Bn] = myFourierCoeff(f, n);
    y = y + An*cos(n*x) + Bn*sin(n*x);
end

plot(x,f(x),x,y)
title(sprintf('%d th Order Fourier Approximation', N))
xlabel('x')
ylabel('y')
legend(`analytic', `approximate')
grid on
axis equal
```

```
>> f = @(x) sin(exp(x));
>> [An, Bn] = myFourierCoeff(f, 3)
An =
    0.109017080509184
Bn =
    0.197555197774232
```

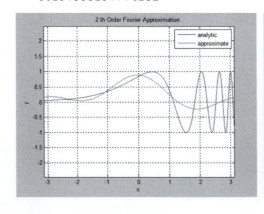
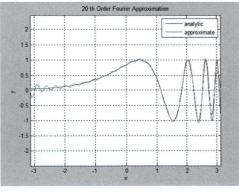

```
>> f = @(x) mod(x,pi/2);
>> [An, Bn] = myFourierCoeff(f, 3)
An =
    6.536814794267476e-007
Bn =
    -3.571571878698915e-006
```

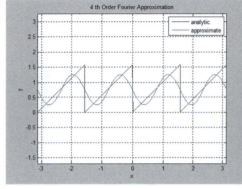
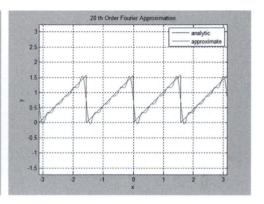

```
>> f = @(x) x > -pi/2 & x < pi/2;
>> [An, Bn] = myFourierCoeff(f, 3)
An =
    -0.212206132273863
Bn =
    0
```

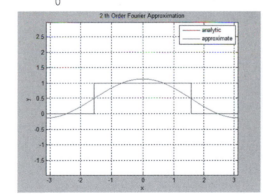
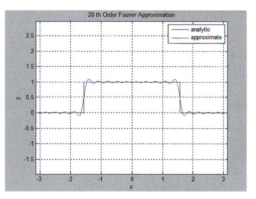

6. For a numerical grid with spacing h, Boole's Rule for approximating integrals says that

$$\int_{x_i}^{x_{i+4}} f(x)dx \approx \frac{3h}{90} \left[7f(x_i) + 32f(x_{i+1}) + 12f(x_{i+2}) + 32f(x_{i+3}) + 7f(x_{i+4}) \right].$$

Show that Boole's Rule is $O(h^7)$ over a single subinterval.

Ordinary Differential Equations (ODEs)

CHAPTER OUTLINE

Motivation

Differential equations are relationships between a function and its derivatives, and they are used to model systems in *every* engineering field. For example, a simple differential equation relates the acceleration of a car with its position. Unlike differentiation where analytic solutions can usually be computed, in general finding exact solutions to differential equations is very hard. Therefore, numerical solutions are critical to making these equations useful for designing and understanding engineering systems.

Because differential equations are so common in engineering, physics, and mathematics, the study of them is a vast and rich field that cannot be covered in this introductory text. This chapter covers ordinary differential equations with specified initial values, a subclass of differential equation problems called initial value problems. To reflect the importance of this class of problem, MATLAB has a whole suite of built-in functions to solve this kind of problem. By the end of this chapter, you should understand what ordinary differential equation initial value problems are, how to pose these problems to MATLAB, and how these MATLAB solvers work.

19.1 ODE Initial Value Problem Statement

A **differential equation** is a relationship between a function, $f(x)$, its independent variable, x, and any number of its derivatives. An **ordinary differential equation** or **ODE** is a differential equation where the independent variable, and therefore also the derivatives, is in one dimension. For the purpose of this

An Introduction to MATLAB® Programming and Numerical Methods. http://dx.doi.org/10.1016/B978-0-12-420228-3.00019-1

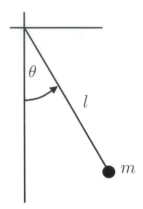

FIGURE 19.1

Pendulum system.

book, we assume that an ODE can be written

$$F\left(x, f(x), \frac{df(x)}{dx}, \frac{d^2 f(x)}{dx^2}, \frac{d^3 f(x)}{dx^3}, \ldots, \frac{d^{n-1} f(x)}{dx^{n-1}}\right) = \frac{d^n f(x)}{dx^n},$$

where F is an arbitrary function that incorporates one or all of the input arguments, and n is the **order** of the differential equation. This equation is referred to as an nth **order ODE**.

To give an example of an ODE, consider a pendulum of length l with a mass, m, at its end; see Figure 19.1. The angle the pendulum makes with the vertical axis over time, $\Theta(t)$, in the presence of vertical gravity, g, can be described by the pendulum equation, which is the ODE

$$ml\frac{d^2\Theta(t)}{dt^2} = -mg\sin(\Theta(t)).$$

This equation can be derived by summing the forces in the x and y direction, and then changing to polar coordinates.

In contrast, a **partial differential equation** or **PDE** is a general form differential equation where x is a vector containing the independent variables $x_1, x_2, x_3, \ldots, x_m$, and the partial derivatives can be of any order and with respect to any combination of variables. An example of a PDE is the heat equation, which describes the evolution of temperature in space over time:

$$\frac{\partial u(t, x, y, z)}{\partial t} = \alpha\left(\frac{\partial u(t, x, y, z)}{\partial x} + \frac{\partial u(t, x, y, z)}{\partial y} + \frac{\partial u(t, x, y, z)}{\partial z}\right).$$

Here, $u(t, x, y, z)$ is the temperature at (x, y, z) at time t, and α is a thermal diffusion constant.

A **general solution** to a differential equation is a $g(x)$ that satisfies the differential equation. Although there are usually many solutions to a differential equation, they are still hard to find. For an ODE of order n, a **particular solution** is a $p(x)$ that satisfies the differential equation *and* n explicitly **known values** of the solution, or its derivatives, at certain points. Generally stated, $p(x)$ must satisfy the differential equation and $p^{(j)}(x_i) = p_i$, where $p^{(j)}$ is the jth derivative of p, for n triplets, (j, x_i, p_i). For the purpose of this text, we refer to the particular solution simply as the **solution**.

TRY IT! Returning to the pendulum example, if we assume the angles are very small (i.e., $\sin(\Theta(t)) \approx \Theta(t)$), then the pendulum equation reduces to

$$l \frac{d^2 \Theta(t)}{dt^2} = -g\Theta(t).$$

Verify that $\Theta(t) = \cos\left(\sqrt{\frac{g}{l}}t\right)$ is a general solution to the pendulum equation. If the angle and angular velocities at $t = 0$ are the known values, Θ_0 and 0, respectively, verify that $\Theta(t) = \Theta_0 \cos\left(\sqrt{\frac{g}{l}}t\right)$ is a particular solution for these known values.

For the general solution, the derivatives of $\Theta(t)$ are

$$\frac{d\Theta(t)}{dt} = -\sqrt{\frac{g}{l}} \sin\left(\sqrt{\frac{g}{l}}t\right)$$

and

$$\frac{d^2 \Theta(t)}{dt^2} = -\frac{g}{l} \cos\left(\sqrt{\frac{g}{l}}t\right).$$

By plugging the second derivative back into the differential equation on the left side, it is easy to verify that $\Theta(t)$ satisfies the equation and so is a general solution.

For the particular solution, the Θ_0 coefficient will carry through the derivatives, and it can be verified that the equation is satisfied. $\Theta(0) = \Theta_0 \cos(0) = \Theta_0$, and $0 = -\Theta_0 \sqrt{\frac{g}{l}} \sin(0) = 0$, therefore the particular solution also satisfies the known values.

A pendulum swinging at small angles is a very uninteresting pendulum indeed. Unfortunately, there is no explicit solution for the pendulum equation with large angles that is as simple algebraically. Since this system is much simpler than most practical engineering systems and has no obvious solution, the need for numerical solutions to ODEs is clear.

A common set of known values for an ODE solution is the **initial value**. For an ODE of order n, the initial value is a known value for the 0th to $n - 1^{\text{th}}$ derivatives at $x = 0$, $f(0)$, $f^{(1)}(0)$, $f^{(2)}(0)$, $\ldots$, $f^{(n-1)}(0)$. For a certain class of ordinary differential equations, the initial value is sufficient to find a unique particular solution. Finding a solution to an ODE given an initial value is called the **initial value problem**. Although the name suggests we will only cover ODEs that evolve in time, initial value problems can also include systems that evolve in other dimensions such as space. Intuitively, the pendulum equation can be solved as an initial value problem because under only the force of gravity, an initial position and velocity should be sufficient to describe the motion of the pendulum for all time afterward.

The remainder of this chapter covers several methods of numerically approximating the solution to initial value problems on a numerical grid. Although initial value problems encompass more than just differential equations in time, we use time as the independent variable. We also use several notations for the derivative of $f(t)$: $f'(t)$, $f^{(1)}(t)$, $\frac{df(t)}{dt}$, and $\dot{f}$, whichever is most convenient for the context.

19.2 Reduction of Order

Many numerical methods for solving initial value problems are designed specifically to solve first-order differential equations. To make these solvers useful for solving higher order differential equations, we must often **reduce the order** of the differential equation to first order. To reduce the order of a differential equation, consider a vector, $S(t)$, which is the **state** of the system as a function of time. In general, the state of a system is a collection of all the dependent variables that are relevant to the behavior of the system. Recalling that the ODEs of interest in this book can be expressed as

$$f^{(n)}(t) = F\left(t, f(t), f^{(1)}(t), f^{(2)}(t), f^{(3)}(t), \ldots, f^{(n-1)}(t)\right),$$

for initial value problems, it is useful to take the state to be

$$S(t) = \begin{bmatrix} f(t) \\ f^{(1)}(t) \\ f^{(2)}(t) \\ f^{(3)}(t) \\ \ldots \\ f^{(n-1)}(t) \end{bmatrix}.$$

Then the derivative of the state is

$$\frac{dS(t)}{dt} = \begin{bmatrix} f^{(1)}(t) \\ f^{(2)}(t) \\ f^{(3)}(t) \\ f^{(4)}(t) \\ \ldots \\ f^{(n)}(t) \end{bmatrix} = \begin{bmatrix} f^{(1)}(t) \\ f^{(2)}(t) \\ f^{(3)}(t) \\ f^{(4)}(t) \\ \ldots \\ F\left(t, f(t), f^{(1)}(t), \ldots, f^{(n-1)}(t)\right) \end{bmatrix} = \begin{bmatrix} S_2(t) \\ S_3(t) \\ S_4(t) \\ S_5(t) \\ \ldots \\ F\left(t, S_1(t), S_2(t), \ldots, S_{n-1}(t)\right) \end{bmatrix},$$

where $S_i(t)$ is the ith element of $S(t)$. With the state written in this way, $\frac{dS(t)}{dt}$ can be written using only $S(t)$ (i.e., no $f(t)$ or its derivatives). In particular, $\frac{dS(t)}{dt} = \mathcal{F}(t, S(t))$, where $\mathcal{F}$ is a function that appropriately assembles the vector describing the derivative of the state. This equation is in the form of a first-order differential equation in S. Essentially, what we have done is turn an nth order ODE into n first order ODEs that are **coupled** together, meaning they share the same terms.

TRY IT! Reduce the second order pendulum equation to first order, where

$$S(t) = \begin{bmatrix} \Theta(t) \\ \dot{\Theta}(t) \end{bmatrix}.$$

Taking the derivative of $S(t)$ and substituting gives the correct expression.

$$\frac{dS(t)}{dt} = \begin{bmatrix} S_2(t) \\ -\frac{g}{l}S_1(t) \end{bmatrix}$$

It happens that this ODE can be written in matrix form:

$$\frac{dS(t)}{dt} = \begin{bmatrix} 0 & 1 \\ -\frac{g}{l} & 0 \end{bmatrix} S(t)$$

ODEs that can be written in this way are said to be **linear ODEs**.

Although reducing the order of an ODE to first order results in an ODE with multiple variables, all the derivatives are still taken with respect to the same independent variable, t. Therefore, the ordinariness of the differential equation is retained.

It is worth noting that the state can hold multiple dependent variables and their derivatives as long as the derivatives are with respect to the same independent variable.

TRY IT! A very simple model to describe the change in population of rabbits, $r(t)$, and wolves, $w(t)$, might be

$$\frac{dr(t)}{dt} = 4r(t) - 2w(t)$$

and

$$\frac{dw(t)}{dt} = r(t) + w(t).$$

The first ODE says that at each time step, the rabbit population multiplies by 4, but each wolf eats two of the rabbits. The second ODE says that at each time step, the population of wolves increases by the number of rabbits and wolves in the system. Write this system of differential equations as an equivalent differential equation in $S(t)$ where

$$S(t) = \begin{bmatrix} r(t) \\ w(t) \end{bmatrix}.$$

The following first-order ODE is equivalent to the pair of ODEs.

$$\frac{dS(t)}{dt} = \begin{bmatrix} 4 & -2 \\ 1 & 1 \end{bmatrix} S(t).$$

19.3 The Euler Method for Solving ODEs

Let $\frac{dS(t)}{dt} = F(t, S(t))$ be an explicitly defined first order ODE. That is, F is a function that returns the derivative, or change, of a state given a time and state value. Also, let t be a numerical grid of the interval $[t_0, t_f]$ with spacing h. Without loss of generality, we assume that $t_0 = 0$, and that $t_f = Nh$ for some positive integer, N.

The linear approximation of $S(t)$ around t_j at t_{j+1} is

$$S(t_{j+1}) = S(t_j) + (t_{j+1} - t_j)\frac{dS(t_j)}{dt},$$

which can also be written

$$S(t_{j+1}) = S(t_j) + hF(t_j, S(t_j)).$$

This formula is called the **Explicit Euler Formula**, and it allows us to compute an approximation for the state at $S(t_{j+1})$ given the state at $S(t_j)$. Starting from a given initial value of $S_0 = S(t_0)$, we can use this formula to integrate the states up to $S(t_f)$; these $S(t)$ values are then an approximation for the solution of the differential equation. The Explicit Euler formula is the simplest and most intuitive method for solving initial value problems. At any state $(t_j, S(t_j))$ it uses F at that state to "point" toward the next state and then moves in that direction a distance of h. Although there are more sophisticated and accurate methods for solving these problems, they all have the same fundamental structure. As such, we enumerate explicitly the steps for solving an initial value problem using the Explicit Euler formula.

WHAT IS HAPPENING? Assume we are given a function $F(t, S(t))$ that computes $\frac{dS(t)}{dt}$, a numerical grid, t, of the interval, $[t_0, t_f]$, and an initial state value $S_0 = S(t_0)$. We can compute $S(t_j)$ for every t_j in t using the following steps.

1. Store $S_0 = S(t_0)$ in an array, S.
2. Compute $S(t_1) = S_0 + hF(t_0, S_0)$.
3. Store $S_1 = S(t_1)$ in S.
4. Compute $S(t_2) = S_1 + hF(t_1, S_1)$.
5. Store $S_2 = S(t_1)$ in S.
6. $\cdots$
7. Compute $S(t_f) = S_{f-1} + hF(t_{f-1}, S_{f-1})$.
8. Store $S_f = S(t_f)$ in S.
9. S is an approximation of the solution to the initial value problem.

When using a method with this structure, we say the method **integrates** the solution of the ODE.

TRY IT! The differential equation $\frac{df(t)}{dt} = e^{-t}$ with initial condition $f_0 = -1$ has the exact solution $f(t) = -e^{-t}$. Approximate the solution to this initial value problem between 0 and 1 in increments of 0.1 using the Explicity Euler Formula. Plot the difference between the approximated solution and the exact solution (Figure 19.2).

```
%% Euler Approximation For Exponential Function
clc; clear all; close all;
```

```
%% Define Parameters
F = @(t,S) exp(-t); % define differential equation
h = 0.1; % define step size
t = 0:h:1; % define numerical grid
f0 = -1; % define initial condition

%% Do Explicit Euler
f = f0; % initialize solution

% for each time step j
for j = 1:length(t)-1
    f(j+1) = f(j) + h*F(t(j),f(j)); % Euler Step
end

%% Plot Approximate and Exact Solution
plot(t, f, 'b--', t, -exp(-t), 'g')
title('Approximate and Exact Solution for Simple ODE')
xlabel('t')
ylabel('f(t)')
grid on
axis tight
legend('Approximate', 'Exact')
```

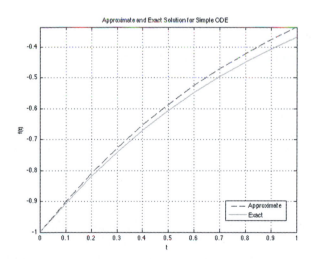

FIGURE 19.2

Comparison of the approximate integration of the function $\frac{df(t)}{dt} = e^{-t}$ between 0 and 1 (dashed) and the exact integration (solid) using Euler's Explicit Formula.

If we repeat the process for $h = 0.01$, we get a better approximation for the solution (Figure 19.3).

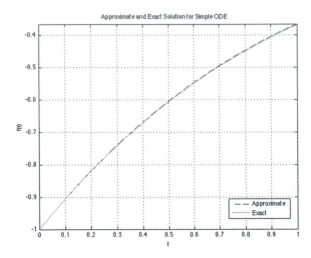

FIGURE 19.3

Comparison of the approximate integration of the function $\frac{df(t)}{dt} = e^{-t}$ between 0 and 1 (dashed) and the exact integration (solid) for smaller step size, h.

The Explicit Euler Formula is called "explicit" because it only requires information at t_j to compute the state at t_{j+1}. That is, $S(t_{j+1})$ can be written explicitly in terms of values we have (i.e., t_j and $S(t_j)$). The **Implicit Euler Formula** can be derived by taking the linear approximation of $S(t)$ around t_{j+1} and computing it at t_j:

$$S(t_{j+1}) = S(t_j) + hF(t_{j+1}, S(t_{j+1})).$$

This formula is peculiar because it requires that we know $S(t_{j+1})$ to compute $S(t_{j+1})$! However, it happens that sometimes we *can* use this formula to approximate the solution to initial value problems. Before we give details on how to solve these problems using the Implicit Euler Formula, we give another implicit formula called the **Trapezoidal Formula**, which is the average of the Explicit and Implicit Euler Formulas:

$$S(t_{j+1}) = S(t_j) + \frac{h}{2}(F(t_j, S(t_j)) + F(t_{j+1}, S(t_{j+1}))).$$

To illustrate how to solve these implicit schemes, consider again the pendulum equation, which has been reduced to first order.

$$\frac{dS(t)}{dt} = \begin{bmatrix} 0 & 1 \\ -\frac{g}{l} & 0 \end{bmatrix} S(t)$$

For this equation,

$$F(t_j, S(t_j)) = \begin{bmatrix} 0 & 1 \\ -\frac{g}{l} & 0 \end{bmatrix} S(t_j).$$

If we plug this expression into the Explicit Euler Formula, we get the following equation:

$$S(t_{j+1}) = S(t_j) + h \begin{bmatrix} 0 & 1 \\ -\frac{g}{l} & 0 \end{bmatrix} S(t_j) = \begin{bmatrix} 1 & 0 \\ 0 & 1 \end{bmatrix} S(t_j) + h \begin{bmatrix} 0 & 1 \\ -\frac{g}{l} & 0 \end{bmatrix} S(t_j) = \begin{bmatrix} 1 & h \\ -\frac{gh}{l} & 1 \end{bmatrix} S(t_j)$$

Similarly, we can plug the same expression into the Implicit Euler to get

$$\begin{bmatrix} 1 & -h \\ \frac{gh}{l} & 1 \end{bmatrix} S(t_{j+1}) = S(t_j),$$

and into the Trapezoidal Formula to get

$$\begin{bmatrix} 1 & -\frac{h}{2} \\ \frac{gh}{2l} & 1 \end{bmatrix} S(t_{j+1}) = \begin{bmatrix} 1 & \frac{h}{2} \\ -\frac{gh}{2l} & 1 \end{bmatrix} S(t_j).$$

With some rearrangement, these equations become, respectively,

$$S(t_{j+1}) = \begin{bmatrix} 1 & -h \\ \frac{gh}{l} & 1 \end{bmatrix}^{-1} S(t_j),$$

$$S(t_{j+1}) = \begin{bmatrix} 1 & -\frac{h}{2} \\ \frac{gh}{2l} & 1 \end{bmatrix}^{-1} \begin{bmatrix} 1 & \frac{h}{2} \\ -\frac{gh}{2l} & 1 \end{bmatrix} S(t_j).$$

These equations allow us to solve the initial value problem, since at each state, $S(t_j)$, we can compute the next state at $S(t_{j+1})$. In general, this is possible to do when an ODE is linear.

19.4 Numerical Error and Instability

There are two main issues to consider with regard to integration schemes for ODEs: **accuracy** and **stability**. Accuracy refers to a scheme's ability to get close to the exact solution, which is usually unknown, as a function of the step size h. Previous chapters have referred to accuracy using the notation $O(h^p)$. The same notation translates to solving ODEs. The stability of an integration scheme is its ability to keep the error from growing as it integrates forward in time. If the error does not grow, then the scheme is stable; otherwise it is unstable. Some integration schemes are stable for certain choices of h and unstable for others; these integration schemes are also referred to as unstable.

To illustrate issues of stability, we numerically solve the pendulum equation using the Euler Explicit, Euler Implicit, and Trapezoidal Formulas.

TRY IT! Use the Euler Explicit, Euler Implicit, and Trapezoidal Formulas to solve the pendulum equation over the time interval $[0, 5]$ in increments of 0.1 and for an initial solution of $S_0 = \begin{bmatrix} 1 \\ 0 \end{bmatrix}$. For the model parameters using $\sqrt{\frac{g}{l}} = 4$. Plot the approximate solution on a single graph (Figure 19.4).

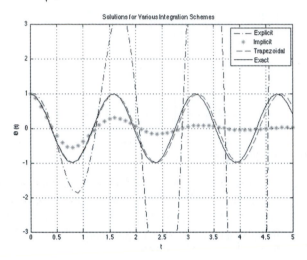

FIGURE 19.4

Comparison of numerical solution to the pendulum problem. The exact solution is a pure cosine wave. The Explicit Euler scheme is clearly unstable. The Implicit Euler scheme decays exponentially, which is not correct. The Trapezoidal method captures the solution correctly, with a small phase shift as time increases.

```
clc;
clear all;
close all;

%% Define Parameters
h       = 0.1;  % define step size
t       = 0:h:5;% define numerical grid
w       = 4;    % oscillation frequency of pendulum
S0      = [1;0];

M_E = [1, h; -w^2*h, 1];
M_I = inv([1, -h; w^2*h, 1]);
M_T = inv([1, -h/2; w^2*h/2, 1])*[1, h/2; -w^2*h/2 1];

%% Do Integrations
S_E = S0';
S_I = S0';
S_T = S0';
```

```
for j = 1:length(t)-1
    S_E(j+1,:) = [M_E*S_E(j,:)']';
    S_I(j+1,:) = [M_I*S_I(j,:)']';
    S_T(j+1,:) = [M_T*S_T(j,:)']';
end

%% Plot Position Solutions
plot(t, S_E(:,1),'b-.', t, S_I(:,1),'g:', t, S_T(:,1),'r—', t,cos(w*t), 'k')
axis([0 5 -3 3])
grid on
title('Solutions for Various Integration Schemes')
xlabel('t')
ylabel('\Theta (t)')
```

19.5 Predictor-Corrector Methods

Given any time and state value, the function, $F(t, S(t))$, returns the change of state $\frac{dS(t)}{dt}$. **Predictor-corrector** methods of solving initial value problems improve the approximation accuracy of non-predictor-corrector methods by querying the F function several times at different locations (predictions), and then using a weighted average of the results (corrections) to update the state.

The **midpoint method** method has a predictor step:

$$S\left(t_j + \frac{h}{2}\right) = S(t_j) + \frac{h}{2}F(t_j, S(t_j)),$$

which is the prediction of the solution value halfway between t_j and t_{j+1}.

It then computes the corrector step:

$$S(t_{j+1}) = S(t_j) + hF\left(t_j + \frac{h}{2}, S\left(t_j + \frac{h}{2}\right)\right),$$

which computes the solution at $S(t_{j+1})$ from $S(t_j)$ but using the derivative from $S\left(t_j + \frac{h}{2}\right)$.

A classical method for integrating ODEs with a high order of accuracy is the **Fourth Order Runge Kutta** (RK4) method. This method uses four predictor corrector steps called k_1, k_2, k_3, and k_4. A weighted average of these predictions is used to produce the approximation of the solution. The formula is as follows.

$$k_1 = F(t_j, S(t_j))$$
$$k_2 = F\left(t_j + \frac{h}{2}, S(t_j) + \frac{1}{2}k_1h\right)$$
$$k_3 = F\left(t_j + \frac{h}{2}, S(t_j) + \frac{1}{2}k_2h\right)$$
$$k_4 = F(t_j + h, S(t_j) + k_3h)$$

The correction step is then

$$S(t_{j+1}) = S(t_j) + \frac{h}{6}\left(k_1 + 2k_2 + 2k_3 + k_4\right).$$

As indicated by its name, the RK4 method is fourth-order accurate, or $O(h^4)$.

19.6 MATLAB® ODE Solvers

MATLAB has several built-in functions for solving initial value problems. These functions all have the same basic construction.

CONSTRUCTION: Initial value problem solvers in MATLAB all have the same basic construction. Let F be a function handle to the function that computes $\frac{dS(t)}{dt} = F(t, S(t))$, t be an array containing the endpoints of the interval of interest, [t0, tf], and S0 be an initial value for S. The function F *must* have the form [dS] = F(t, S), although the name does not have to be F. The input argument t is assumed to be a scalar, and the input argument S is assumed to be a column vector the same size as S0.

   If the function ode is an arbitrary ODE solver, it can be called using the following syntax:

   [T,S]=ode(F, t, S0)

   Here, T is a column vector of the integrated time values, and S is an array of the solution approximations for each element of the state. S will have as many rows as the length of T, and as many columns as the length of S0. Row j of S is the approximation of the solution, $S(t_j)$. In other words, S(j,:) is the approximation of solution at T(j).

WARNING! The input arguments for F must have both t and S, even if one of them is not used to compute the derivative.

   The most commonly used ODE solver is ode45. We illustrate the use of ode45 with two simple examples and then a more interesting example.

EXAMPLE: Consider the ODE

$$\frac{dS(t)}{dt} = \cos(t)$$

for an initial value $S_0 = 0$. The exact solution to this problem is $S(t) = \sin(t)$. Use ode45 to approximate the solution to this initial value problem over the interval $[0, \pi]$. Plot the approximate solution versus the exact solution and the relative error over time.

```
F = @(t,S) cos(t);
[T,S]=ode45(F,[0,pi],0);
plot(T,S,'LineWidth',2);
grid on
axis tight
xlabel('t')
ylabel('S(t)')
title('Approximation to Solution of Simple ODE')
figure
plot(T,S-sin(T),'LineWidth',2)
grid on
axis tight
xlabel('t')
ylabel('S(t) - sin(t)')
title('Approximation Error from ODE45')
```

As can be seen from Figure 19.5, the difference between the approximate and exact solution to this ODE is extremely small.

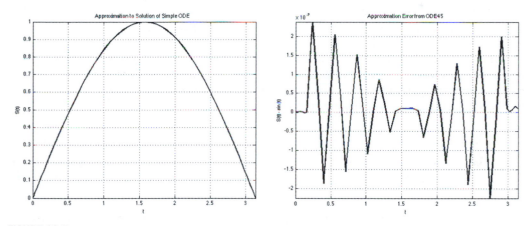

FIGURE 19.5

Left: Integration of $\frac{dS(t)}{dt} = \cos(t)$ with ode45. **Right:** Computation of the difference between the solution of the integration by ode45 and the evaluation of the analytical solution to this ODE.

TIP! In the previous example, we used anonymous function handles to produce the input function for ode45. It is useful to store more complicated input functions as m-files and create handles to them using the @ symbol.

EXAMPLE: Consider the ODE

$$\frac{dS(t)}{dt} = -S(t),$$

with an initial value of $S_0 = 1$. The exact solution to this problem is $S(t) = e^{-t}$. Use ode45 to approximate the solution to this initial value problem over the interval [0, 1]. Plot the approximate solution versus the exact solution, and the relative error over time.

```
F = @(t,S) -S;
[T,S]=ode45(F,[0,1],1);
plot(T,S,'LineWidth',2)
xlabel('t')
ylabel('S(t)')
title('Approximation to Solution of Simple ODE')
grid on
axis tight
figure
plot(T,S-exp(-T),'LineWidth',2)
grid on
xlabel('t')
ylabel('S(t) - exp(-t)')
title('Approximation Error from ODE45')
```

Figure 19.6 shows the corresponding numerical results. As in the previous example, the difference between the result of ode45 and the evaluation of the analytical solution by MATLAB is very small in comparison to the value of the function.

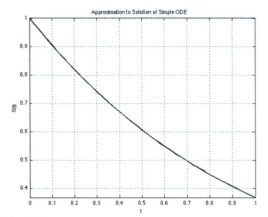 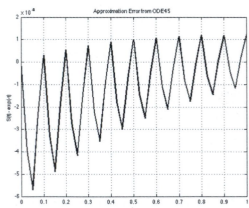

FIGURE 19.6

Left: Integration of $\frac{dS(t)}{dt} = -y(t)$ with ode45. **Right:** Computation of the difference between the solution of the integration by ode45 and the evaluation of the analytical solution to this ODE.

TRY IT! Let the state of a system be defined by $S(t) = \begin{bmatrix} x(t) \\ y(t) \end{bmatrix}$, and let the evolution of the system be defined by the ODE

$$\frac{dS(t)}{dt} = \begin{bmatrix} 0 & t^2 \\ -t & 0 \end{bmatrix} S(t).$$

Use `ode45` to solve this ODE for the time interval [0, 10] with an initial value of $S_0 = \begin{bmatrix} 1 \\ 1 \end{bmatrix}$. Plot the solution in $(x(t), y(t))$ (see Figure 19.7).

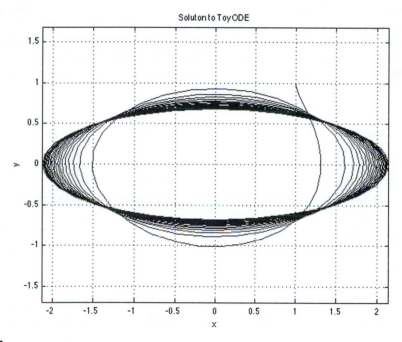

FIGURE 19.7

Solution to Toy ODE using `ode45`.

```
>> F = @(t, S) [0 t^2; -t 0]*S;
>> [T,S] = ode45(F, [0 10], [1 ; 1]);
>> plot(S(:,1), S(:,2))
>> axis tight
>> axis equal
>> grid on
>> xlabel('x')
>> ylabel('y')
>> title('Soluton to Toy ODE')
```

Table 19.1   Description of MATLAB's ODE Solvers

Solver	Problem Type	Order of Accuracy	When to Use
ode45	Nonstiff	Medium	Most of the time. This should be the first solver you try.
ode23	Nonstiff	Low	For problems with crude error tolerances or for solving moderately stiff problems.
ode113	Nonstiff	Low to high	For problems with stringent error tolerances or for solving computationally intensive problems.
ode15s	Stiff	Low to medium	If ode45 is slow because the problem is stiff.
ode23s	Stiff	Low	If using crude error tolerances to solve stiff systems and the mass matrix is constant.
ode23t	Moderately Stiff	Low	For moderately stiff problems if you need a solution without numerical damping.
ode23tb	Stiff	Low	If using crude error tolerances to solve stiff systems.

In practice, some ODEs have bad behavior known as **stiffness**. Loosely speaking, stiffness refers to systems that can have very sharp changes in derivative. An example of a stiff system is a bouncing ball, which suddenly changes directions when it hits the ground. Depending on the properties of the ODE you are solving and the desired level of accuracy, you might need to use different solvers. Table 19.1 shows some of MATLAB's ODE solvers, their accuracy, and the stiffness conditions in which they can be used.

Summary

1. Ordinary differential equations (ODEs) are equations that relate a function to its derivatives, and initial value problems are a specific kind of ODE solving problem.
2. Most initial value problems cannot be integrated explicitly and therefore require numerical solutions.
3. There are explicit, implicit, and predictor-corrector methods for numerically solving initial value problems.
4. The accuracy of the scheme used depends on its order of approximation of the ODE.
5. The stability of the scheme used depends on the ODE, the scheme, and the choice of the integration parameters.

Vocabulary

accuracy
coupled
differential equation
Explicit Euler Formula
Fourth-Order Runge Kutta Method
general solution

Midpoint Method
n^{th} order ODE
ODE
ordinary differential equation
partial differential equation
particular solution

solution
stability
state
Trapezoidal Formula

Implicit Euler Formula	PDE
initial value problem	predictor corrector schemes
linear ODE	reduce the order

Functions and Operators

ode15s	ode23t	ode113
ode23	ode23tb	
ode23s	ode45	

Problems

⌞m⌟ **1.** The logistics equation is a simple differential equation model that can be used to relate the change in population dP/dt to the current population, P, given a growth rate, r, and a carrying capacity, K. The logistics equation can be expressed by:

$$\frac{dP}{dt} = rP\left(1 - \frac{P}{K}\right)$$

Write a function with header [dP] = myLogisticsEq(t, P, r, K) that represents the logistics equation. Note that this format allows myLogisticsEq to be used as an input argument to ode45. You may assume that the arguments dP, t, P, r, and K are all scalars, and dP is the value $\frac{dP}{dt}$ given r, P, and K. Note that the input argument, t, is obligatory if myLogisticsEq is to be used as an input argument to ode45, even though it is part of the differential equation.

Note: The logistics equation has an analytic solution defined by:

$$P(t) = \frac{K P_0 e^{rt}}{K + P_0(e^{rt} - 1)}$$

where P_0 is the initial population. As an exercise, you should verify that this equation is a solution to the logistics equation.

Test Cases:

```
>> dP = myLogisticsEq(0, 10, 1.1, 15)
dP =
    3.6667
>> t0 = 0; tf = 20; P0 = 10; r = 1.1; K = 20; t = 0:.01:20;
>> [T, P] = ode45(@myLogisticsEq, [t0 tf], P0, [], r, K);
>> plot(T, P, t, K*P0*exp(r*t)./(K + P0*(exp(r*t) - 1)))
>> title('Numerical and Analytic Solution of Logistic Equation')
>> xlabel('time')
>> ylabel('population')
>> legend('Numerical Solution', 'Exact Solution')
>> grid on
>> axis tight
```

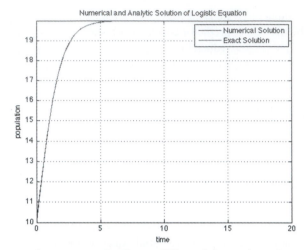

Numerical and Analytic Solution of Logistic Equation

.m 2a. The Lorenz attractor is a system of ordinary differential equations that was originally developed to model convection currents in the atmosphere. The Lorenz equations can be written as:

$$\frac{dx}{dt} = \sigma(y - x)$$

$$\frac{dy}{dt} = x(\rho - z) - y$$

$$\frac{dz}{dt} = xy - \beta z$$

where x, y, and z represent position in three dimensions and σ, ρ, and β are scalar parameters of the system. You can read more about the Lorenz attractor on Wikipedia: http://en.wikipedia.org/wiki/Lorenz_equation.

Write a function with header [dS] = myLorenz(t,S,sigma,rho,beta), where t is a scalar denoting time, S is a 3×1 array denoting the position (x, y, z), and sigma, rho, and beta are strictly positive scalars representing σ, ρ, and β. The output argument dS should be the same size as S.

Test Cases:
```
>> dS = myLorenz(0, [1; 2; 3], 10, 28, 8/3)
dS =
      10
      23
      -6
```

.m 2b. Write a function with header [T, X, Y, Z] = myLorenzSolver(tSpan, s0, sigma, rho, beta) that solves the Lorenz equations using ode45. The input argument tSpan should be a 1×2 array of the form [t0, tf], where t0 is the initial time, and tf is the final time of consideration. The input argument s0 should be a 3×1 array of the form [x0; y0; z0], where (x_0, y_0, z_0) represents an initial position. Finally, the input arguments sigma, rho, and beta are the scalar parameters σ, ρ, and β of the Lorenz system. The output arguments

T should be an array of times given as the output argument of ode45. The output arguments, X, Y, and Z should be the numerically integrated solution produced from myLorenz and ode45.

Note: Your function, myLorenz.m, from problem 2a should be a subfunction of myLorenzSolver. You do not need to submit myLorenz.m.

Test Cases:

```
sigma = 10; rho = 28; beta = 8/3;
t0 = 0; tf = 50;
S0 = [0 1 1.05];
[T, X, Y, Z] = myLorenzSolver([t0, tf], S0, sigma, rho, beta);
plot3(X,Y,Z,'b')
title(['Lorenz Attractor Solution for \sigma = ',num2str(sigma),', \rho = ', num2str(rho),
    ', \beta = ', num2str(beta)])
xlabel('x'); ylabel('y'); zlabel('z')
grid on
```

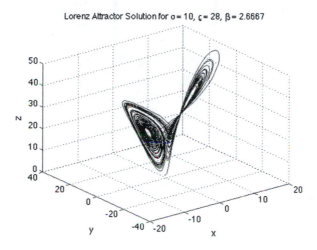

```
for an animated test case:
for i = 1:length(T)
    plot3(X(1:i),Y(1:i),Z(1:i),'b', X(i), Y(i), Z(i), 'ro')
    title('Lorenz Attractor')
    xlabel('x'); ylabel('y'); zlabel('z')
    axis([-30 30 -30 30 0 50])
    view([45,45])
    grid on
    drawnow
end
```

Special Test Case: Go to   http://www.mathworks.com/matlabcentral/fileexchange/31133-plot3anianaglyph and download the file plot3AniAnaglyph.m from MATLAB Central by clicking 'download all' in the upper left-hand corner. For future reference, this website is a good repository of functions that other people have built. Who knows? You may find that someone has already solved your problem for you!

Place the function into the working directory. You can call the help for `plot3AniAnaglyph` to learn how it works. After running `myLorenzSolver`, the following line of code will produce a 3D animation of the Lorenz attractor. Be sure to have those 3D glasses on!

```
>> plot3AniAnaglyph(X,Y,Z)
```

.m 3. Consider the following model of a mass-spring-damper (MSD) system in one dimension. In this figure m denotes the mass of the block, c is called the damping coefficient, and k is the spring stiffness. A damper is a mechanism that dissipates energy in the system by resisting velocity. The MSD system is a simplistic model of several engineering applications such as shock observers and structural systems.

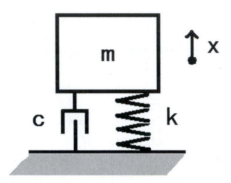

The relationship between acceleration, velocity, and displacement can be expressed by the following mass-spring-damper (MSD) differential equation:

$$m\ddot{x} + c\dot{x} + kx = 0$$

which can be rewritten:

$$\ddot{x} = \frac{-(c\dot{x} + kx)}{m}$$

Let the state of the system be denoted by the vector $S = [x; v]$ where x is the displacement of the mass from its resting configuration and v is its velocity. Rewrite the MSD equation as a first-order differential equation in terms of the state, S. In other words, rewrite the MSD equation as $dS/dt = f(t, S)$.

Write a function with header `[dS] = myMSD(t, S, m, c, k)`, where `t` is a scalar denoting time, `S` is a 2×1 vector denoting the state of the MSD system, and `m`, `c`, and `k` are the mass, damping, and stiffness coefficients of the MSD equation, respectively.

Test Cases:

```
>> dS = myMSD(0, [1; −1], 10, 1, 100)
dS =
    −1.0000
    −9.9000
```

```
>> m = 1; k = 10;
>> [T0, S0] = ode45(@myMSD, [0 20], [1 0], [], m, 0, k);
>> [T1, S1] = ode45(@myMSD, [0 20], [1 0], [], m, 1, k);
>> [T2, S2] = ode45(@myMSD, [0 20], [1 0], [], m, 10, k);
>> plot(T0,S0(:,1), T1, S1(:,1), T2, S2(:,1))
>> title('Numerical Solution of MSD System with Varying Damping')
>> xlabel('time')
>> ylabel('displacement')
>> legend('no damping', 'c = 1', '> critcally damped')
```

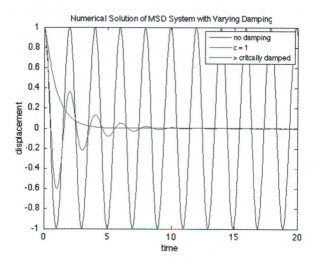

4. Write a function with header [T, S] = myForwardEuler(dS), tSpan, S0, where dS is a handle to a function, $f(t, S)$, describing a first-order differential equation, tSpan is an array of times for which numerical solutions of the differential equation are desired, and S0 is the initial condition of the system. Assume that the size of the state is one. The output argument, T, should be a column vector such that T(i) = tSpan(i) for all i, and S should be the integrated values of dS at times T. You should perform the integration using the Forward Euler method, $S(t_i) = S(t_{i-1}) + (t_i \cdot t_{i-1})dS(t_{i-1}, S(t_{i-1}))$.

  Note: $S(1)$ should equal $S0$.

```
>> dS = @(t, S) t*exp(-S);
>> tSpan = linspace(0,1,10);
>> S0 = 1;
>> [Teul, Seul] = myForwardEuler(dS, tSpan, S0);
>> Teul'
ans =
0 0.1111 0.2222 0.3333 0.4444 0.5556 0.6667 0.7778 0.8889 1.0000
>> Seul'
ans =
1.0000 1.0000 1.0045 1.0136 1.0270 1.0447 1.0664 1.0919 1.1209 1.1531
```

```
>> [Tode, Sode] = ode45(dS, tSpan, S0);
>> t = linspace(0,1,1000);
>> plot(Teul, Seul, Tode, Sode, t, log(exp(S0) + (t.^2 - t(1)^2)/2)) ?= check !!
>> title('Numerical Solutions of an ODE')
>> xlabel('t')
>> ylabel('S(t)')
>> legend('Euler', 'ode45', 'exact')
```

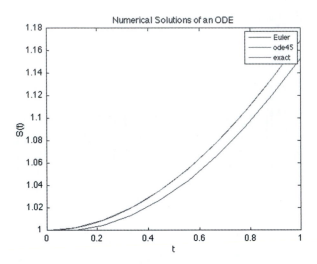

5. Write a function with header [T, S] = myRK4(dS, tSpan, S0), where the input and output arguments are the same as in problem 4. The function myRK4 should numerically integrate dS using the fourth-order Runge-Kutta method.

Test Cases:

```
>> dS = @(t, S) sin(exp(S))/(t+1);
>> tSpan = linspace(0,2*pi,10);
>> S0 = 0;
>> [Trk, Srk] = myRK4(dS, tSpan, S0);
>> [Tode, Sode] = ode45(dS, tSpan, S0);
>> plot(Trk, Srk, Tode, Sode)
>> title('Numerical Solutions of an ODE')
>> xlabel('t')
>> ylabel('S(t)')
>> legend('RK4', 'ode45')
```

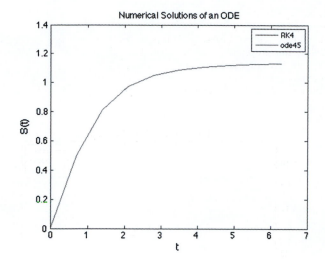

Index